Reality Is Not What It Seems

ABOUT THE AUTHOR

Carlo Rovelli is a theoretical physicist who has made significant contributions to the physics of space and time. He has worked in Italy and the US, and is currently directing the quantum gravity research group of the Centre de Physique Théorique in Marseille, France. His *Seven Brief Lessons on Physics* is a phenomenal international bestseller translated into forty-one languages.

Reality Is Not What It Seems

It Seems

The Journey to Quantum Gravity

CARLO ROVELLI

Translated by
Simon Carnell and Erica Segre

ALLEN LANE
an imprint of
PENGUIN BOOKS

ALLEN LANE

UK | USA | Canada | Ireland | Australia
India | New Zealand | South Africa

Allen Lane is part of the Penguin Random House group of companies
whose addresses can be found at global.penguinrandomhouse.com

Penguin
Random House
UK

First published in Italian under the title *La realtà non è come ci appare* by Raffaello Cortina Editore SpA 2104
This translation first published in Great Britain by Allen Lane 2016

Set in 12.5/15 pt Garamond MT Std
Typeset by Jouve (UK), Milton Keynes
Printed in Great Britain by Clays Ltd, St Ives plc

A CIP catalogue record for this book is available from the British Library

ISBN: 978-0-241-25796-8

www.greenpenguin.co.uk

MIX
Paper from
responsible sources
FSC
www.fsc.org   FSC® C018179

Penguin Random House is committed to a
sustainable future for our business, our readers
and our planet. This book is made from Forest
Stewardship Council® certified paper.

Contents

PART FOUR

Beyond Space and Time

Author's Note

During my entire research life, friends and curious people have asked me to explain what was going on in quantum gravity research. How was it possible to study new ways of thinking about space and time? Over and over again I have been asked to write a popular account of this research. While books on cosmology or string theory abound, a book describing the research on the quantum nature of space and time, and on loop quantum gravity in particular, did not yet exist. I have long hesitated, because I wanted to concentrate on research. Some years ago, after completing my technical book on the subject, I felt that the collective work of many scientists had moved the topic to a stage mature enough for a popular book. The landscape we are exploring is enchanting: why keep it hidden?

But I still delayed the project, because I could not 'see' the book in my head. How to explain a world without space and time? One night in 2012, during a long solitary drive from Italy to France, I realized that the only way to explain in a comprehensible manner the ongoing modifications of the notions of space and time was to tell the story from the beginning: starting from Democritus, all the way through to the quanta of space. After all, this is how I understand the story. I began to design the entire book in my mind while driving, and got increasingly excited, until I heard a police car's sirens telling me to pull over: I was driving far above the speed limit. The Italian policemen asked me politely if I was crazy to drive at that speed. I explained that I had just found the idea I'd been seeking for so long; the policeman let me go without a ticket, and wished me good luck with the book. This is the book.

This book was written and first published in Italian at the beginning of 2014. Shortly afterwards, I wrote a few articles on fundamental physics for an Italian newspaper. A prestigious Italian publisher, Adelphi, asked me for an extended version of these articles, to appear as a small booklet. This is the origin of the short book *Seven Brief Lessons on Physics*, which to my immense surprise has become an international bestseller and has opened a beautiful channel of communication between me and so many wonderful readers all over the world. The *Seven Lessons* were thus written *after* this book, and to some extent they are a synthesis of some of the topics you find here. If you have read *Seven Brief Lessons on Physics* and want to know more, to journey deeper into the strange world that book sketched, here you can find more.

While the account of established physics I give here is presented from the peculiar perspective in which I understand it, it is largely uncontroversial. However, the part of this book that describes current research in quantum gravity is my own personal understanding of the state of the art. This is the region at the boundary between what we have understood and what we do not yet understand, and is still far from achieving consensus. Some of my physicist colleagues will agree with what I write here; others won't. This is true for all presentations of ongoing research at the frontiers of knowledge, but I prefer to state it upfront and clearly. This is not a book about certainties: it is a book about the adventure of moving towards the unknown.

As a whole, this is a travel book describing one of the

most spectacular journeys that humanity has taken: a journey out of our limited and parochial views of reality, towards an increasingly vast understanding of the structure of things. A magical journey out of our common-sense view of things, far from complete.

Marseille, 4 May 2016

Preface: Walking along the Shore

We are obsessed with ourselves. We study *our* history, *our* psychology, *our* philosophy, *our* gods. Much of our knowledge revolves around man himself, as if we were the most important thing in the universe. I think I like physics because it opens a window through which we can see further. It gives me the sense of fresh air entering the house.

What we see out there through the window is constantly surprising us. We have learned a great deal about the universe. In the course of the centuries we have come to realize just how very many wrong ideas we had. We thought that the Earth was flat, and that it was the still centre of our world. That the universe was small, and unchanging. We believed that man was a breed apart, without kinship to the other animals. We have learned of the existence of quarks, black holes, particles of light, waves of space, and of the extraordinary molecular structures in every cell of our bodies. The human race is like a growing child who discovers with amazement that the world consists not just of his bedroom and playground, but that it is vast, and that there are a thousand things to discover, and innumerable ideas quite different from those with which he began. The universe is multiform and boundless, and we continue to stumble upon new aspects of it. The more we learn about the world, the more we are amazed by its variety, beauty and simplicity.

But the more we discover, the more we understand that what we don't yet know is greater than what we know. The more powerful our telescopes, the stranger and more unexpected are the heavens we see. The closer we look at the minute detail of matter, the more we discover of its profound structure. Today we see almost to the Big Bang, the great explosion from which, 14 billion years ago, all the galaxies were born – but we have already begun to glimpse something beyond the Big Bang. We have learned that space is curved, but already foresee that this same space is woven from vibrating quantum grains.

Our knowledge of the elementary grammar of the world continues to grow. If we try to put together what we have learned about the physical world in the course of the twentieth century, the clues point towards something profoundly different from what we were taught at school. An elementary structure of the world is emerging, generated by a swarm of quantum events, where time and space do not exist. Quantum fields draw space, time, matter and light, exchanging information between one event and another. Reality is a network of granular events; the dynamic which connects them is probabilistic; between one event and another, space, time, matter and energy melt in a cloud of probability.

This strange new world is slowly emerging today from the study of the main open question posed in fundamental physics: quantum gravity. It's the problem of coherently synthesizing what we have learned about the world with the two major discoveries of twentieth-century physics: general relativity and quantum theory. To quantum gravity, and the strange world that this research is unfolding, this book is dedicated.

The book is a live coverage of the ongoing research: what we are learning, what we already know, and what we think we are beginning to understand, about the elementary nature of things. It starts from the distant origin of some key ideas that we use today to order our understanding of the world and describes the two great discoveries of the twentieth century – Einstein's general relativity and quantum mechanics – trying to put into focus the core of their physical content. It tells of the picture of the world which is emerging today from research in quantum gravity, taking into account the latest indications given by nature, such as the confirmation of the cosmological standard model obtained from the Planck satellite and the failure at CERN to observe the super-symmetric particles that many expected. And it discusses the consequences of these ideas: the granular structure of space; the disappearance of time at small scale; the physics of the Big Bang; the origin of black-hole heat – up to the role of information in the foundation of physics.

In a famous myth related by Plato in the seventh book of *The Republic*, some men are chained at the bottom of a dark cave and see only shadows cast upon a wall by a fire behind them. They think that this is reality. One of them frees himself, leaves the cave and discovers the light of the Sun, and the wider world. At first the light, to which his eyes are unaccustomed, stuns and confuses him. But eventually he can see, and returns excitedly to his companions to tell them what he has seen. They find it hard to believe.

We are all in the depths of a cave, chained by our ignorance, by our prejudices, and our weak senses reveal to us only shadows. If we try to see further, we are confused: we are unaccustomed. But we try. This is science. Scientific thinking explores and redraws the world, gradually offering us better and better images of it, teaching us to think in ever more effective ways. Science is a continual exploration of ways of thinking. Its strength is its visionary capacity to demolish preconceived ideas, to reveal new regions of reality, and to construct novel and more effective images of the world. This adventure rests upon the entirety of past knowledge, but at its heart is change. The world is boundless and iridescent; we want to go and see it. We are immersed in its mystery and in its beauty, and over the horizon there is unexplored territory. The incompleteness and the uncertainty of our knowledge, our precariousness, suspended over the abyss of the immensity of what we don't know, does not render life meaningless: it makes it interesting and precious.

I have written this book to give an account of what for me is the wonder of this adventure. I've written with a particular reader in mind: someone who knows little or nothing about today's physics but is curious to find out what we know, but also what we don't yet understand, about the elementary weave of the world – and where we are searching. And I have written it to try to communicate the breathtaking beauty of the panorama of reality which can be seen from this perspective.

I've also written it for my colleagues, fellow travellers dispersed throughout the world, as well as for the young women and men with a passion for science, eager to set out on this journey for the first time. I've sought to outline the general landscape of the structure of the physical world, as seen by the double lights of relativity and of quantum physics, and to show how they can be combined. This is not only a book of divulgation; it's also one which articulates a point of view, in a field of research where the abstraction of the technical language may sometimes obscure the wide-angle vision. Science is made up of experiments, hypotheses, equations, calculations and long discussions; but these are only tools, like the instruments of musicians. In the end, what matters in music is the music itself, and what matters in science is the under-standing of the world which science provides. To understand the significance of the discovery that the Earth turns around the Sun, it is not necessary to follow Copernicus's compli-cated calculations; to understand the importance of the discovery that all living beings on our planet have the same ancestors, it is not necessary to follow the complex argu-ments of Darwin's books. Science is about reading the world from a gradually widening point of view.

This book gives an account of the current state of the search for our new image of the world, as I understand it today. It is the reply I would give to a colleague and friend asking me, 'So, what do you think is the true nature of things?', as we walk along the shore, on a long midsummer's evening.

PART ONE
Roots

This book begins in Miletus, twenty-six centuries ago. Why begin a book about quantum gravity with events, people and ideas so ancient? I hope the reader eager to get on to quanta of space will not hold this against me. For it is easier to understand ideas by starting with the roots from which they have grown, and an important number of the ideas which turned out to be effective for understanding the world originated over two thousand years ago. If we briefly retrace their birth, they become clearer, and the later steps turn out to be simpler and natural.

But there's more. Certain problems first posed in antiquity continue to be crucial to our understanding of the world. Some of the most recent ideas about the structure of space utilize concepts and issues introduced then. In speaking of these distant ideas, I put on to the table questions which are going to be central to quantum gravity. This makes it also possible, when treating of quantum gravity, to distinguish between the ideas which go back to the very origin of scientific thought, even if we are unfamiliar with them, and those which are radically new. The connection between problems posed by the scientists of antiquity, and solutions found by Einstein and quantum gravity, is, as we shall see, surprisingly close.

1. Grains

According to tradition, in the year 450 BCE, a man embarked upon a ship travelling from Miletus to Abdera. It was to be a crucial journey for the history of knowledge.

The man was probably fleeing political turmoil in Miletus, where the aristocracy was violently seizing back power. Miletus had been a prosperous and flourishing Greek city, perhaps the principal city of the Greek world before the golden age of Athens and Sparta. It had been a busy commercial hub, dominating a network of almost a hundred colonies and commercial outposts, stretching from the Black Sea to Egypt.

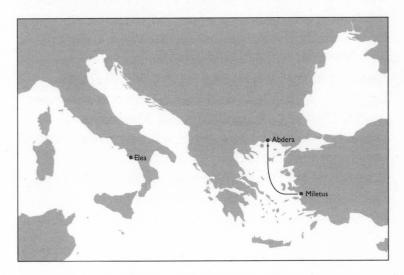

Figure 1.1 The journey made by Leucippus of Miletus, the founder of the atomist school (circa 450 BCE).

Caravans from Mesopotamia and ships from all over the Mediterranean arrived at Miletus, and ideas circulated.

During the preceding century, a revolution in thinking which would prove fundamental to humanity had taken place in Miletus. A group of thinkers had reformulated the way questions were asked about the world, and the way answers were sought. The greatest of these thinkers was Anaximander.

From time immemorial, or at least since humanity had left written texts which have come down to us, men had asked themselves how the world had come into being, what it was composed of, how it was ordered, and why natural phenomena occurred. For thousands of years they had given themselves answers which all resembled one another: answers which referred to elaborate stories of spirits, deities, imaginary and mythological creatures, and other similar things. From cuneiform tablets to ancient Chinese texts; from hieroglyphic writing in the Pyramids to the myths of the Sioux; from the most ancient Indian texts to the Bible; from African stories to those of aboriginal Australians, it was all a colourful but basically quite monotonous flow – of Plumed Serpents and Great Cows, of irascible, litigious, or kindly deities who create the world by breathing over abysses, uttering '*Fiat lux*', or emerging out of a stone egg.

Then, at Miletus, at the beginning of the fifth century before our era, Thales, his pupil Anaximander, Hecataeus and their school find a different way of looking for answers. This immense revolution in thought inaugurates a new mode of knowledge and understanding, and signals the first dawn of scientific thought.

The Milesians understand that by shrewdly using observation and reason, rather than searching for answers in fantasy, ancient myths or religion – and, above all, by using critical thought in a discriminating way – it is possible to repeatedly correct our world view, and to discover new aspects of reality which are hidden to the common view. It is possible to discover the new.

Perhaps the decisive discovery is that of a different style of thinking, where the disciple is no longer obliged to respect and share the ideas of the master but is free to build on those ideas without being afraid to discard or criticize the part that can be improved. This is a novel middle way, placed between full adherence to a school and generic deprecation of ideas. It is the key to the subsequent development of philosophical and scientific thinking: from this moment onwards, knowledge begins to grow at a vertiginous pace, nourished by past knowledge but at the same time by the possibility of criticism, and therefore of improving knowledge and understanding. The dazzling *incipit* of Hecataeus's book of history goes to the heart of this critical thinking, including as it does the awareness of our own fallibility: 'I wrote things which seem true to me, because the accounts of the Greeks seem to be full of contradictory and ridiculous things.'

According to legend, Heracles descended to Hades from Cape Tenaro. Hecataeus visits Cape Tenaro, and determines that there is in fact no subterranean passage or other access to Hades there – and therefore judges the legend to be false. This marks the dawn of a new era.

This new approach to knowledge works quickly and impressively. Within a matter of a few years, Anaximander understands that the Earth floats in the sky and the sky continues beneath the Earth; that rainwater comes from the evaporation of water on Earth; that the variety of substances in the world must be susceptible to being understood in terms of a single, unitary and simple constituent, which he calls *apeiron*, the indistinct; that the animals and plants evolve and adapt to changes in the environment, and that man must have evolved from other animals. Thus, gradually, was founded the basis of a grammar for understanding the world which is substantially still our own today.

Situated at a point of conjunction between the emergent Greek civilization and the ancient empires of Mesopotamia and Egypt, nourished by their knowledge but immersed in the liberty and the political fluidity which is typically Greek; in a social space without imperial palaces, or powerful priestly castes, where individual citizens discuss their destinies in open agoras, Miletus is the place where, for the first time, men decide collectively their own laws; where the first parliament in the history of the world gathers – the *Panionium*, meeting-place of the delegates of the Ionian League – and where for the first time men doubt that only the gods are capable of accounting for the mysteries of the world. Through discussion, it is possible to reach the best decisions for the community; through discussion, it is possible to understand the world. This is the immense legacy of Miletus, cradle of philosophy, of the natural sciences, and of geographical and historical studies. It is no exaggeration to say that the entire scientific and philosophical tradition, Mediterranean and then modern, has a crucial root in the speculations of the thinkers of Miletus in the sixth century BCE.[1]

This luminous Miletus shortly afterwards came to a calamitous end. The arrival of the Persian Empire, and a failed anti-imperial revolt, led to a ferocious destruction of the city in 494 BCE and to the enslavement of a large number of its inhabitants. In Athens, the poet Phrynichus writes the tragedy *The Taking of Miletus*, which so deeply moves the Athenians that its re-staging is prohibited, on account of it causing too much distress. But twenty years later, the Greeks repel the Persian menace; Miletus is reborn, repopulated, and returned to being a centre of commerce and ideas, radiating its thought and spirit once more.

The person with whom we began this chapter must have been moved by this spirit when, in 450, according to tradition, he embarked from Miletus for Abdera. His name was Leucippus. Little is known about his life.[2] He wrote a book entitled *The Great Cosmology*. On his arrival in Abdera, he founded a scientific and philosophical school to which he soon affiliated a young disciple, Democritus, whose long shadow was to be cast over the thought of all subsequent times.

Together, these two thinkers have built the majestic cathedral of ancient atomism. Leucippus was the teacher. Democritus, the great pupil who wrote dozens of works on every field of knowledge, and was deeply venerated in antiquity, when people were familiar with these works. 'The most subtle of the Ancients', Seneca called him.[3] 'Who is there whom we can compare with him for the greatness, not merely of his genius, but also of his spirit?' asks Cicero.[4]

What, then, had Leucippus and Democritus discovered? The Milesians had understood that the world can be comprehended using reason. They had become convinced that the variety of natural phenomena must be attributable to something simple, and had tried

Figure 1.2 Democritus of Abdera.

7

to understand what this something might be. They had conceived of a kind of elementary substance of which everything was made. Anaximenes, among the Milesians, had imagined this substance could compress and rarefy, thus transforming from one to another of the elements of which the world is constituted. It was a first germ of physics, rough and elementary, but in the right direction. An idea was needed, a great idea, a grand vision, to grasp the hidden order of the world. Leucippus and Democritus came up with this idea.

The idea of Democritus's system is extremely simple: the entire universe is made up of a boundless space in which innumerable atoms run. Space is without limits; has neither an above nor a below; is without a centre, or a boundary. Atoms have no qualities at all, apart from their shape. They have no weight, no colour, no taste. 'By convention is sweet, by convention bitter, by convention hot, by convention colour; but by verity atoms and void.'[5]

Atoms are indivisible; they are the elementary grains of reality, which cannot be further subdivided, and everything is made of them. They move freely in space, colliding one with another; they hook on to and push and pull each other. Similar atoms attract each other and join.

This is the weave of the world. This is reality. Everything else is nothing but a by-product, random and accidental, of this movement and this combining of atoms. The infinite variety of the substances of which the world is made derives solely from this combining of atoms.

When atoms aggregate, the only thing that matters, the only thing that exists at the elementary level, is their shape, their arrangement, and the order in which they combine. Just as by combining the letters of the alphabet in different ways we may obtain comedies or tragedies, ridiculous stories or epic poems,

8

so elementary atoms combine to produce the world in its endless variety. The metaphor is Democritus's own.[6]

There is no finality, no purpose, in this endless dance of atoms. We, just like the rest of the natural world, are one of the many products of this infinite dance. The product, that is, of an accidental combination. Nature continues to experiment with forms and structures; and we, like the animals, are the products of a selection which is random and accidental, over the course of eons of time. Our life is a combination of atoms, our thoughts are made up of thin atoms, our dreams are the products of atoms; our hopes and our emotions are written in a language formed by combinations of atoms; the light which we see is comprised of atoms which bring us images. The seas are made of atoms, as are our cities, and the stars. It's an immense vision; boundless, incredibly simple and incredibly powerful, one on which the knowledge of a civilization would later be built.

On this foundation Democritus wrote dozens of books articulating a vast system, dealing with questions of physics, philosophy, ethics, politics and cosmology. He writes on the nature of language, on religion, on the origins of human societies, and on much else besides. (The opening of his *Little Cosmology* is impressive: 'In this work I treat of all things.') All of these books have been lost. We know of his thought only through the quotations and references made by other ancient authors, and by their summaries of his ideas.[7] The thought which thus emerges is a kind of intense humanism, rationalist and materialist.[8] Democritus combines a keen attention to nature, illuminated by a naturalistic clarity in which every residual system of mythic ideas is cleared away, with a great attention to humanity and a deep ethical concern for life – anticipating by some two thousand years the best aspects of the eighteenth-century Enlightenment. The ethical ideal of Democritus is that of a serenity of mind reached through moderation and balance, by trusting in reason and not allowing oneself to be overwhelmed by passions.

Plato and Aristotle were familiar with Democritus's ideas, and fought against them. They did so on behalf of other ideas, some of which were later, for centuries, to create obstacles to the growth of knowledge. Both insisted on rejecting Democritus's naturalistic explanations, in favour of trying to understand the world in finalistic terms – believing, that is, that everything that happens has a purpose; a way of thinking that would reveal itself to be very misleading for understanding the ways of nature – or in terms of good and evil, confusing human issues with matters which do not relate to us.

Aristotle speaks extensively about the ideas of Democritus, and with respect. Plato never cites Democritus, but scholars suspect today that this was out of deliberate choice and not for lack of knowledge of his works. Criticism of Democritus's ideas is implicit in several of Plato's texts, as in his critique of 'physicists', for example. In a passage in his *Phaedo*, Plato has Socrates articulate a reproach to all 'physicists' which will have a lasting resonance. He complains that when 'physicists' had explained that the Earth was round, he rebelled because he wanted to know what 'good' it was for the Earth to be round; how its roundness would benefit it. Plato's Socrates recounts how he had at first been enthusiastic about physics, but had come to be disillusioned by it:

I had expected to be first told that the Earth was flat or round, but also that, afterwards, the reason for the necessity of this shape would be explained to me, starting from the principle of the best, proving to me that the best thing for the Earth is to have this shape. And if he had said that the Earth was at the centre of the world, then to show me how being at the centre was of benefit to the Earth.[9]

How completely off track the great Plato was here!

Is there a limit to divisibility?

The greatest physicist of the second half of the twentieth century, Richard Feynman, wrote at the beginning of his wonderful introductory lessons on physics:

If, in some cataclysm, all scientific knowledge were to be destroyed, and only one sentence passed on to the next generation of creatures, what statement would contain the most information in the fewest words? I believe it is the atomic hypothesis, or the atomic fact, or whatever you wish to call it, that *all things are made of atoms – little particles that move around in perpetual motion, attracting each other when they are a little distance apart, but repelling upon being squeezed into one another.* In that one sentence you will see an enormous amount of information about the world, if just a little imagination and thinking are applied.[10]

Without needing anything from modern physics, Democritus had already arrived at the idea that everything is made up of indivisible particles. How did he do it?

He had arguments based upon observation; for example, he imagined, correctly, that the wearing down of a wheel, or the drying of clothes on a line, could be due to the slow flight of particles of wood or of water. But he also had arguments of a philosophical kind. Let's focus on these, because their potency reaches all the way to quantum gravity.

Democritus observed that matter could not be a continuous whole, because there is something contradictory in the proposition that it should be so. We know of Democritus's reasoning because Aristotle reports it.[11] Imagine, says Democritus, that matter is infinitely divisible, that is to say, it may be broken down an infinite number of times. Imagine then that you break up a piece of matter ad infinitum. What would be left?

Could small particles of extended dimension remain? No, because if this were the case the piece of matter would not yet be broken up to infinity. Therefore, only points without extension would remain. But now let us try to put together the piece of matter starting from these points: by putting together two points without extension you cannot obtain a thing with extension, nor can you with three, or even with four. No matter how many you put together, in fact, you never have extension, because points have no extension. Therefore, we cannot think that matter is made of points without extension, because no matter how many of these we manage to put together, we never obtain something with an extended dimension. The only possibility, Democritus concludes, is that any piece of matter is made up of a *finite* number of discrete pieces which are indivisible, each one having *finite* size: the atoms.

The origin of this subtle mode of argumentation pre-dates Democritus. It comes from the Cilento region in the south of Italy, from a town now called Velia, which in the fifth century BCE was a flourishing Greek colony called Elea. Parmenides lived there, the philosopher who had taken to the letter – perhaps too much – the rationalism of Miletus and the idea born there that reason can reveal to us how things can be other than they appear. Parmenides had explored an avenue to truth via pure reason alone, a path which led him to declare that all appearances are illusory, thus opening a direction of thinking that would progressively move towards metaphysics and distance itself from what would come to be known as 'natural science'. His pupil Zeno, also from Elea, had brought subtle arguments to bear in support of this fundamentalist rationalism, which refutes the credibility of appearances radically. Among these arguments there was a series of paradoxes that became celebrated as 'Zeno's paradoxes'; they seek to show how all appearance is illusory, arguing that the commonplace notion of motion is absurd.[12]

The most famous of Zeno's paradoxes is presented in the form of a brief fable: the tortoise challenges Achilles to a race, starting out with a ten-metre advantage. Will Achilles manage to catch up with the tortoise? Zeno argues that rigorous logic dictates that he will never be able to do so. Before catching up, in effect, Achilles needs to cover the ten metres, and in order to do this he will take a certain amount of time. During this time, the tortoise will have advanced a few centimetres. To cover these centimetres, Achilles will have to take a little more time but, meanwhile, the tortoise will have advanced further, and so on, ad infinitum. Achilles therefore requires *an infinite number of such times* to reach the tortoise, and an *infinite number of times*, argues Zeno, is *an infinite amount of time*. Consequently, according to strict logic, Achilles will take an infinite time to reach the tortoise; or rather, we will never see him do so. Since, however, we do see the swift Achilles reaching and overtaking as many tortoises as he likes, it follows that what we see is irrational, and therefore illusory.

Let's be honest: this is hardly convincing. Where does the error lie? One possible answer is that Zeno is wrong because it is not true that by accumulating an infinite number of things one ends up with an infinite thing. Think of taking a piece of string, cutting it in half, and then again in half, and again, ad infinitum. At the end, you will obtain an infinite number of small pieces of string; the sum of these, however, will be finite, because they can only add up to the length of the original piece of string. Hence, an infinite number of strings can make a finite string; an infinite number of increasingly short times may make a *finite* time, and the hero, even if he will have to cover an infinite number of distances, ever smaller, will take a *finite* time to do so, and will end up catching the tortoise.

It seems that the paradox is resolved. The solution, that is, is in the idea of the continuum – arbitrarily small times may exist, an infinite number of which make up a finite time. Aristotle is the first to intuit this possibility, which was subsequently developed by ancient and modern mathematics.*

But is this really the correct solution in the *real* world? Do

* In technical terms, there are converging infinite sums. For the example of the string, the infinite sum $\frac{1}{2} + \frac{1}{4} + \frac{1}{8} + \frac{1}{16} \ldots$ converges to 1. Infinite convergent sums were not understood in Zeno's time. Archimedes understood them a few centuries later, and used them to calculate areas. Newton used them heavily, but not until the nineteenth century, with Bolzano and Weierstrass, was conceptual clarity on these mathematical objects achieved. Aristotle, however, had already understood that this was a possible way to answer Zeno; the Aristotelian distinction between actual infinity and potential infinity already contains the key idea: the difference between the absence of a limit to divisibility, and the

arbitrarily short strings really exist? Can we really cut a piece of string an *arbitrary* number of times? Do infinitely small amounts of time exist? This is precisely the problem that quantum gravity will have to face.

According to tradition, Zeno had met Leucippus and had become his teacher. Leucippus was therefore familiar with Zeno's riddles. But he had devised a *different* way of resolving them. Maybe, Leucippus suggests, nothing arbitrarily small exists: there is a lower limit to divisibility.

The universe is granular, not continuous. With infinitely small points, it would be impossible ever to construct extension – as in Democritus's argument, reported by Aristotle and mentioned above. Therefore, the extension of the string must be formed by a *finite* number of finite objects with *finite* size. The string *cannot* be cut *as many times as we want*; matter is not continuous, it is made of individual 'atoms' of a finite size.

Whether this abstract argument is correct or not, its conclusion – as we know today – contains a great deal of truth. Matter does indeed have an atomic structure. If I divide a drop of water in two, I obtain two drops of water. I can divide each one of these two drops again, and so on. But I cannot continue to infinity. At a certain point I have only one molecule, and I have finished. No drops of water exist smaller than a single molecule of water.

How do we know this today? Evidence has accumulated over centuries, much of it from chemistry. Chemical substances are made up of combinations of a few elements and are formed by proportions (of weight) given by whole

_____

possibility of having already divided something an infinite number of times.

numbers. Chemists have constructed a way of thinking about substances as composed of molecules made up of fixed combinations of atoms. Water, for example – H_2O – is composed of two parts hydrogen and one part oxygen.

But these are only clues. Still at the beginning of the last century numerous scientists and philosophers did not consider the atomic hypothesis to be credible. Among them was the renowned physicist and philosopher Ernst Mach, whose ideas on space would come to have great importance for Einstein. At the end of a lecture by Ludwig Boltzmann at the Imperial Academy of Science in Vienna, Mach publicly declared, 'I do not believe that atoms exist!' This was in 1897. Many, like Mach, understood chemical notation only as a conventional method of summarizing laws of chemical reactions – not as evidence that there actually were molecules of water composed of two atoms of hydrogen and one of oxygen. You can't see atoms, they would say. Atoms will never be seen, they would say. And then, they asked, how big would an atom be? Democritus could never measure the size of his atoms . . .

But somebody else could. The definitive proof of the 'atomic hypothesis' had to wait until 1905. It was found by a rebellious twenty-five-year-old, who had studied physics but had not been able to find employment as a scientist and was making ends meet by working as an employee in the patent office in Berne. I will speak a lot about this young man in the rest of this book, and about the three articles he sent to the most prestigious physics journal of the time, the *Annalen der Physik*. The first of these articles contained the definitive proof that atoms exist, and calculated their dimensions, solving the problem posed by Leucippus and Democritus twenty-three centuries earlier.

The name of this twenty-five-year-old, obviously, is Albert Einstein.

How does he do it? The idea is surprisingly simple. Anyone could have arrived at it, from the time of Democritus onwards, if he had had Einstein's acumen and a sufficient mastery of mathematics to make what was not an easy calculation. The idea goes like this: if we observe attentively very small particles, such as a speck of dust or a grain of pollen, suspended in still air or in a liquid, we see them tremble and dance. Pushed by this trembling,

Figure 1.3 Albert Einstein.

they move, randomly zigzagging, and so they drift slowly, gradually moving away from their starting point. This motion of particles in a fluid is called Brownian motion, after Robert Brown, a biologist who described it in detail in the nineteenth century. The typical trajectory of a particle dancing in this manner is illustrated in figure 1.4. It is as if the small particle is receiving blows randomly from each side of it. In fact, it isn't 'as if' it were being hit, it really is hit. It trembles because it is hit by individual molecules of air, which collide with the particle at times from the right and at times from the left.

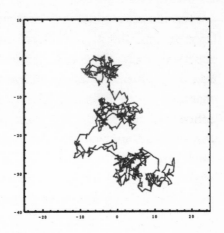

Figure 1.4 Typical Brownian motion.

The subtle point is the following. There are an enormous number of molecules of air. On average, as many hit the granule from the left as those that hit it from the right. If the air's molecules were infinitely small and infinitely numerous, the effect of the collisions from right and from left would balance and thus cancel out at each instant, and the granule would not move. But the finite size of the molecules – the fact that these are present in *finite* rather than infinite number – causes there to be *fluctuations* (this is the key word): that is to say, the collisions never balance out *exactly*; they only balance out *on average*. Imagine for a moment the molecules were very few in number and large in size: the granule would clearly receive a blow only occasionally; now one on the right, then one on the left . . . Between one collision and the other it would move here and there to a significant degree, like a football kicked by boys running around a playing field. The smaller the molecules, on the other hand, the shorter the interval between collisions and the better the hits from different directions balance and cancel each other out. And the less the granule moves.

It is indeed possible, with a little mathematics, to work back from the amount of movement of the granule, which can be observed, to the dimensions of the molecules. Einstein does this, as I mentioned above, at the age of twenty-five. From observations of granules drifting in fluids, from the measurement of how much these 'drift' – that is, move away from a position – he calculates the dimensions of Democritus's atoms, the elementary grains of which matter is made. He provides, after 2,300 years, the proof of the accuracy of Democritus's insight: matter is granular.

The nature of things

Sublime Lucretius's work will not die,
Until the day the world itself passes away.
— Ovid[13]

I often think that the loss of the works of Democritus in their entirety* is the greatest intellectual tragedy to ensue from the collapse of the old classical civilization. Take a look at the list of his works in the footnote; it is difficult not to be dismayed, imagining what we have lost of the vast scientific reflections of antiquity.

We have been left with all of Aristotle, by way of which

---

* Here is the list of all of the works of Democritus, with their titles as given by Diogenes Laertius: *Great Cosmology; Little Cosmology; Cosmography; On the Planets; On Nature; On Human Nature; On Intelligence; On the Senses; On the Soul; On Flavours; On Colour; On Diverse Movements of the Atoms; Of Changes in Shape; The Causes of Celestial Phenomena; The Causes of Atmospheric Phenomena; On Fire and On Things in Fire; The Causes of Acoustic Phenomena; Concerning the Magnet; The Causes of Seeds, Plants and Fruits; On Animals; A Description of the Sky; Geography; A Description of the Pole; On Geometry; Geometrical Reality; On the Tangents of the Circle and the Sphere; Numbers; On Irrational Lines and Solids; Projections; Astronomy; Astronomical Table; On Rays of Light; On Reflected Images; On Rhythm and Harmony; On Poetry; On the Beauty of Song; On Euphony and Cacophony; Concerning Homer, or on Correct Epic Diction; The Science of Medicine; On Agriculture; On Words; On Names; On Values, or On Virtue; On the Disposition which Characterizes the Wise; On Painting; A Treatise on Tactics; Circumnavigation of the Ocean; On History; The Thought of Chaldea; The Thought of the Phrygians; On the Sacred Writings of Babylon; On the Sacred Writings of Meroe; On Fevers and the Coughs Deriving from Illness; On Aporiae; Legal Questions; Pythagoras; On Logic, or Criterion of Thought; Confirmations; Points of Ethics; On Well-being.* All lost . . .

Western thought reconstructed itself, and nothing by Democritus. Perhaps, if all of the works of Democritus had survived, and nothing of Aristotle's, the intellectual history of our civilization would have been better . . .

But centuries dominated by monotheism have not permitted the survival of Democritus's naturalism. The closure of the ancient schools such as those of Athens and Alexandria and the destruction of all the texts not in accordance with Christian ideas were vast and systematic, at the time of the brutal anti-pagan repression following from the edicts of Emperor Theodosius, which, in 390–1 declared that Christianity was to be the only and obligatory religion of the empire. Plato and Aristotle, pagans who believed in the immortality of the soul or in the existence of a Prime Mover, could be tolerated by a triumphant Christianity. Not Democritus.

But a text survived the disaster, and has reached us in its entirety. Through it, we know a little about ancient atomism and, above all, we know the spirit of that science. It is the splendid poem *De rerum natura* (*The Nature of Things*, or *On the Nature of the Universe*), by the Latin poet Lucretius.

Lucretius adheres to the philosophy of Epicurus, a pupil of a pupil of Democritus. Epicurus is interested more in ethical than scientific questions and does not have Democritus's depth. He sometimes translates Democritean atomism a little superficially. But his vision of the natural world is substantially that of the great philosopher of Abdera. Lucretius decants in verse the thought of Epicurus and the atomism of Democritus, and in this way a part of this profound philosophy was saved from the intellectual catastrophe of the Dark Ages. Lucretius sings of atoms, the sea, the sky, of nature. He expresses in luminous verse philosophical questions, scientific ideas, refined arguments.

. . . I will explain by what forces nature steers the courses of the sun and the journeyings of the moon, so that we shall not suppose that they run their yearly races between the heaven and earth of their own free will . . . or that they are rolled round in furtherance of some divine plan . . .[14]

The beauty of the poem lies in the sense of wonder which pervades the vast atomistic vision – the sense of the profound unity of things, derived from the knowledge that we are all made of the same substance as are the stars, and the sea:

. . . we are all sprung from heavenly seed. All alike have the same father, from whom all-nourishing mother earth receives the showering drops of moisture. Thus fertilized, she gives birth to smiling crops and lusty trees, to mankind and all the breeds of beasts. She it is that yields the food on which they all feed their bodies, lead their joyous lives and renew their race . . .[15]

There is a sense of luminous calm and serenity about the poem, which comes from understanding that there are no capricious gods demanding of us difficult things, and punishing us. There is a vibrant and airy joyfulness, beginning with the marvellous opening verses dedicated to Venus, a radiant symbol of the creative force of nature:

Before you the winds flee, and at your coming the clouds forsake the sky. For you the ocean levels laugh, the sky is calmed and glows with diffused radiance.[16]

There is a deep acceptance of the life of which we are an integral part:

Do you not see that nature is clamouring for two things only,
a body free from pain, a mind released from worry and fear
for the enjoyment of pleasurable sensations?[17]

And there is a serene acceptance of the inevitability of
death, which cancels every evil and about which there is
nothing to fear. For Lucretius, religion is ignorance; reason is
the torch that brings light.

Lucretius's text, forgotten for centuries, was rediscovered
in January 1417 by the humanist Poggio Bracciolini, in the
library of a German monastery. Poggio had been the secre-
tary of many popes, and was a passionate hunter of ancient
books, in the wake of the celebrated rediscoveries made
by Francesco Petrarch. His rediscovery of a text by Quintil-
ian modified the course of the study of law throughout
the faculties of Europe; his discovery of the treatise on
architecture by Vitruvius transformed the way in which
fine buildings were designed and constructed. But his triumph was rediscovering Lucretius. The actual codex found
by Poggio has been lost, but the copy made by his friend
Niccolò Niccoli (now known as the Codex Laurenziano
35.30) is still preserved in its entirety in Florence's Biblioteca
Laurenziana.

The ground was already surely prepared for something
new when Poggio gave Lucretius's book back to humanity.
Already, from Dante's generation, it had been possible to
hear markedly new accents:

> Your eyes went shooting through my heart
> to wake my dormant thought.
> Look now, I'm desperate and distraught
> with love that tears my life apart.[18]

But the rediscovery of *De rerum natura* had a profound effect upon the Italian and European Renaissance,[19] and its echo resounds, directly or indirectly, in the pages of authors ranging from Galileo[20] to Kepler,[21] and from Bacon to Machiavelli. In Shakespeare, a century after Poggio, atoms make a delightful appearance:

> MERCUTIO O, then I see Queen Mab hath been with you:
>   She is the fairies' midwife, and she comes
>   In shape no bigger than an agate-stone
>   On the forefinger of an alderman,
>   Drawn with a little team of atomies
>   Over men's noses as they lie asleep . . .[22]

Montaigne's *Essays* include at least a hundred quotations from Lucretius. But the direct influence of Lucretius extended to Newton, Dalton, Spinoza, Darwin – and all the way up to Einstein. The very idea of Einstein's that the existence of atoms is revealed by the Brownian motion of minute particles immersed in a fluid may be traced back to Lucretius. Here is a passage in which Lucretius provides a 'living proof' of the notion of atoms:

> This process is illustrated by an image of it that is continually taking place before our very eyes. Observe what happens when sunbeams are admitted into a building and shed light on its shadowy places. You will see a multitude of tiny particles mingling in a multitude of ways in the empty space within the light of the beam, as though contending in everlasting conflict, rushing into battle rank upon rank with never a moment's pause in a rapid sequence of unions and disunions. From this you may picture what it is for the atoms

to be perpetually tossed about in the illimitable void. To some extent a small thing may afford an illustration and an imperfect image of great things. Besides, there is a further reason why you should give your mind to these particles that are seen dancing in a sunbeam: their dancing is an actual indication of underlying movements of matter that are hidden from our sight. There you will see many particles under the impact of invisible blows, changing their course and driven back upon their tracks, this way and that, in all directions. You must understand that they all derive this restlessness from the atoms. It originates with the atoms, which move of themselves. Then those small compound bodies that are least removed from the impetus of the atoms are set in motion by the impact of their invisible blows and in turn cannon against slightly larger bodies. So the movement mounts up from the atoms and gradually emerges to the level of our senses, so that those bodies are in motion that we see in sunbeams, moved by blows that remain invisible.[23]

Einstein resuscitated the 'living proof' presented by Lucretius, and probably first conceived of by Democritus, and made it solid by translating it into mathematical terms, thus managing to calculate the size of the atoms.

The Catholic Church attempted to stop Lucretius: in the Florentine Synod of December 1516 it prohibited the reading of Lucretius in schools. In 1551, the Council of Trent banned his work. But it was too late. An entire vision of the world which had been swept away by medieval Christian fundamentalism was re-emerging in a Europe which had reopened its eyes. It was not just the rationalism, atheism and materialism of Lucretius that were being proposed in

Europe. It was not merely a luminous and serene meditation on the beauty of the world. It was much more: it was an articulate and complex structure of thinking about reality, a new mode of thinking, radically different from what had been for centuries the mind-set of the Middle Ages.[24]

The medieval cosmos so marvellously sung by Dante was interpreted on the basis of a hierarchical organization of the universe which reflected the hierarchical organization of European society: a spherical cosmic structure with the Earth at its centre; the irreducible separation between Earth and heavens; finalistic and metaphorical explanations of natural phenomena. Fear of God, fear of death; little attention to nature; the idea that forms preceding things determine the structure of the world; the idea that the source of knowledge could only be the past, in revelation and tradition . . .

There is none of this in the world of Democritus as sung by Lucretius. There is no fear of the gods; no ends or purposes in the world; no cosmic hierarchy; no distinction between Earth and heavens. There is a deep love of nature, a serene immersion within it; a recognition that we are profoundly part of it; that men, women, animals, plants and clouds are organic threads of a marvellous whole, without hierarchies. There is a feeling of deep universalism in the wake of the splendid words of Democritus: 'To a wise man, the whole earth is open, because the true country of a virtuous soul is the entire universe.'[25]

There is, too, the ambition of being able to think about the world in simple terms. Of being able to investigate and understand the secrets of nature. To know more than our parents. And there are extraordinary conceptual tools on which Galileo, Kepler and Newton will build: the idea of free rectilinear motion in space; the idea of elementary bodies and their interactions, out of which the world is constructed; the idea of space as a container of the world.

And there is the simple idea of the finite divisibility of things. The granular quality of the world. The idea which stops the infinite between our fingers. This idea is at the root of the atomic hypothesis, but it will also return with augmented force with quantum mechanics, and today is revealing itself to be powerful again – as the keystone of quantum gravity.

The first person to make the parts of the mosaic which begin to emerge from Renaissance naturalism cohere – and to reprise the Democritean vision, immensely reinforced, placing it at the centre of modern thought – will be an Englishman, the greatest scientist of all time and the first protagonist of the following chapter.

2. The Classics

Isaac and the little moon

If in the previous chapter I appeared to be saying that Plato and Aristotle have only done harm to the development of science, I would like to correct this impression. Aristotle's studies of nature – of botany and zoology, for example – are extraordinary scientific works, grounded upon meticulous observations of the natural world. The conceptual clarity, the attention to the variety of nature, the impressive intelligence and the openness of mind of the great philosopher made him an authority for centuries to come. The first systematic physics that we know of is Aristotle's, and it is not bad physics at all.

Aristotle presents it in a book entitled, precisely, *Physics*. The book didn't take its title from the name of a discipline: it was the discipline which got its name from Aristotle's book. For Aristotle, physics works as follows. First, it is necessary to distinguish between the heavens and Earth. In the heavens, everything is made up of a crystalline substance which moves in a circular motion and turns eternally around the Earth in great concentric circles, with the spherical Earth at the centre of everything. On the Earth, it is necessary to distinguish between forced motion and natural motion. Forced motion is caused by a thrust and ends when the thrust ends. Natural motion is vertical – upwards or downwards – and depends both on the substance and the location. Each substance has a 'natural place', that is to say, a proper altitude to which it always returns: earth at the bottom, water a little way above it, air a little higher still, and fire even higher. When you pick up a stone and let it fall, the stone moves downwards because it wants to return to its natural level. Air bubbles in water, fire in the air; and children's balloons move upwards, seeking their natural place.

Do not smile at this theory, or dismiss it, because it is very sound physics. It's a good and correct description of the motion of bodies immersed in a fluid and subject to gravity and friction, namely, the real things we meet in our everyday experience. It's not wrong physics, as is frequently said.* It's an approximation. But the physics of Newton, too, is an approximation of general relativity. And probably everything that we know today as well is an approximation of something else which we don't yet know. The physics of Aristotle is still rough, it is not quantitative (we cannot compute with it), but it is coherent and rational and enables correct qualitative predictions to be made. It is not for nothing that it remained for centuries the best available model for understanding motion.[1]

Perhaps even more important for the future development of science is Plato.

It is he who understood the value of the intuition of Pythagoras and Pythagorism: that the key to moving forwards, to going beyond Miletus, was mathematics.

Pythagoras was born on Samos, a small island not far from

* The bad reputation of Aristotelian physics dates back to the polemics of Galileo. Galileo had to move forward and therefore needed to be critical. He attacked Aristotle viciously, with scorn and sarcasm. But he took Aristotle's physics very seriously.

Miletus. His first biographers, Iamblichus and Porphyry, report how young Pythagoras was a disciple of elderly Anaximander. Everything originates in Miletus. Pythagoras travelled widely, probably in Egypt and as far as Babylon, before finally settling in the south of Italy, at Crotone, where he founded a religio-politico-scientific sect which played an important role in the politics of the little town but left a momentous legacy to the entire world: the discovery of the theoretical utility of mathematics: 'Number' – he is said to have asserted – 'governs forms and ideas.'[2]

Plato divested Pythagorism of its cumbersome and useless mystical baggage. He absorbed and distilled its useful message: mathematics is the language best adapted to understand and describe the world. The reach of this insight is immense; it is one of the reasons for the success of Western science. According to tradition, Plato carved on the door of his school the phrase, 'Let no one enter here who is ignorant of geometry.'

Driven by this conviction, it was Plato who posed the momentous question: the question out of which, after a long detour, modern science would emerge. Of his disciples who studied mathematics he asked if they could find the mathematical laws followed by the celestial bodies visible in the heavens. Venus, Mars and Jupiter can be easily observed in the night sky. They seem to move a little at random, back and forth among the other stars. Is it possible to find a mathematics which is able to describe and predict their movements?

The exercise begins with Eudoxus in Plato's school, and is pursued throughout the following centuries by astronomers such as Aristarchus and Hipparchus, bringing ancient astronomy on to an extremely high scientific level. We know of the triumphs of this science thanks to a single book, the only one to have survived: the *Almagest* of Ptolemy. Ptolemy was an astronomer who lived in Alexandria in the first century of our era, under the Roman Empire, when science was already in decline and about to disappear altogether, overwhelmed by the collapse of the Hellenistic world and suffocated by the Christianization of the empire.

Ptolemy's book is a major work of science. Rigorous, precise, complex, it presents a mathematical system of astronomy capable of predicting the seemingly random movements of the planets in the sky, with almost complete precision, given the limitations of the human eye. The book is the proof that the intuition of Pythagoras was correct. Mathematics allows the world to be described and the future to be predicted: the apparently wandering and disorderly movements of the planets can be precisely predicted by using mathematical formulae that Ptolemy, summarizing the results of centuries of work by Greek astronomers, presents in a systematic and masterly way. Even today, with a little studying, it is possible to open Ptolemy's book, learn its techniques and *calculate*, for example, the position which Mars will have in a *future* sky. *Today*: two thousand years, that is, after the book was written. The realization that working this magic is really possible is the basis of modern science and owes not a little to Pythagoras and Plato.

After the collapse of ancient science, no one throughout the Mediterranean was capable of understanding Ptolemy – or any of the other small number of major scientific works that survived the catastrophe, such as the *Elements* of Euclid. In India, where Greek learning had arrived, thanks to rich commercial and cultural exchanges, these books were studied and understood.

From India this knowledge returned to the West, thanks

to learned Persian and Arab scientists who were able to understand and preserve it. But astronomy did not take any very significant step forward for over a thousand years.

More or less at the same time as Poggio Bracciolini discovered the manuscript of Lucretius, the heady atmosphere of Italian humanism and the enthusiasm for ancient texts also intoxicated a young Pole who had come to study in Italy, first at Bologna, then at Padua. He signed himself in the Latin manner: Nicolaus Copernicus. The young Copernicus studies Ptolemy's *Almagest* and falls in love with it. He decides to spend his life doing astronomy, following in the footsteps of the great Ptolemy.

The time is now ripe and, more than a thousand years after Ptolemy, Copernicus is able to make the leap forward that generations of Indian, Arab and Persian astronomers had not been able to make: not simply learning, applying and adding small ameliorations to the Ptolemaic system but thoroughly improving it – with the courage to change it in depth. Instead of describing heavenly bodies turning around the Earth, Copernicus writes a sort of revised and corrected version of Ptolemy's *Almagest*, in which the Sun is at the centre and the Earth, together with the other planets, runs around it.

In this way – Copernicus hopes – the calculations would work even better. In reality, they did not work much better than those of Ptolemy; in fact, in the end, they turned out to work less well. But the idea was, nevertheless, sound: in the succeeding generation, Johannes Kepler shows that the Copernican system can really be made to function better than Ptolemy's. Painstakingly analysing new, precise observations, Kepler shows that a few new mathematical laws can describe with exactitude the movements of the planets around the Sun, with a degree of accuracy even greater than any obtained in antiquity. We are now in 1600, and for the first time humanity finds out how to do something better than what was done in Alexandria more than a thousand years earlier.

While Kepler in the cold north calculates movements in the sky, in Italy it is with Galileo Galilei that the new science begins to take off. Exuberant, Italian, polemical, argumentative, highly cultured, exceptionally intelligent and overflowing with inventiveness, Galileo gets sent from Holland a new invention – the telescope – and makes a gesture which changes human history. He points it towards the sky.

Like Roy in *Blade Runner,* he sees things we people wouldn't believe: rings around Saturn, mountains on the moon, phases of Venus, moons orbiting Jupiter . . . Each of these phenomena renders Copernicus's ideas more plausible. Scientific instruments begin to open the myopic eyes of humankind on to a world vaster and more varied than it had as yet been able to conceive of.

But Galileo's grand idea is to make the logical deduction from the cosmic revolution brought about by Copernicus: convinced that the Earth is a planet like all others, Galileo reasons that, if movements in the heavens follow precise mathematical laws, and if the Earth is a planet like all others, and thus part of the heavens, then there must also exist precise mathematical laws governing the movements of objects *on Earth*.

Confident of the rationality of nature and of the Pythagorean-Platonic vision that nature is understandable through mathematics, Galileo decides to study *how* objects move on Earth when they are set free – that is, when they fall. Convinced that a relevant mathematical law must exist, he sets out to search for it, by trial and error. For the first time in the history of mankind, an *experiment* is made. Experimental science begins with Galileo. His experiment is simple: he lets objects fall; that is, he lets them follow what for Aristotle was their natural movement and seeks to measure precisely their falling speed.

The result is momentous: objects do not fall always at a constant speed, as everybody thought. Their speed, instead, gradually increases during the early part of the fall. In this phase, what is constant is not the speed of the fall but rather the *acceleration*, that is to say, the rate at which speed increases. And, magically, this acceleration turns out to be the same for all objects. Galileo completes a first rough measurement of this acceleration and finds it constant. Its value is approximately 9.8 metres per second per second, which is to say that for every second that an object falls, its speed increases by 9.8 metres per second. Keep in mind this number.

This is the first mathematical law discovered for earthly bodies: the law of falling bodies.* Up until this point, only mathematical laws for the movements of the planets had been discovered. Mathematical perfection is no longer confined to the heavens.

But the greatest result is still to come, and it will be Isaac Newton who will attain it. Newton studies in depth the results of Galileo and Kepler and by combining them, finds the hidden diamond. We can follow his reasoning in terms of the 'little moon', as he himself tells it in *The Mathematical Principles of Natural Philosophy*, the book that forms the foundation of modern science.

Imagine that the Earth, writes Newton, had many moons, like Jupiter. In addition to the real Moon, imagine other

* $(x = \frac{1}{2} \, a \, t^2)$.

moons, and in particular a *little moon* which orbits the Earth at a minimal distance from it, just above the mountain peaks. At what speed would this little moon travel? One of the laws discovered by Kepler relates the radius of orbit with period, namely, the time taken to complete an orbit.* We know the radius of the orbit of the real Moon (Hipparchus had measured it in antiquity) and its period (one month). We know the radius of the orbit of the little moon (the radius of the Earth, measured by Eratosthenes in antiquity). With a simple proportion we can compute the period of the orbit of the little moon. The result is one hour and a half. The little moon would complete its orbit around the Earth every one hour and a half.

Now, an object which orbits does not go straight: it continually changes direction, and a change of direction is an *acceleration*. The little moon accelerates towards the centre of the Earth. This acceleration is easy to compute.† Newton makes the simple calculation, and the result is . . . 9.8 metres per second per second! The same acceleration as in Galileo's experiments for falling bodies on Earth.

Coincidence? It can't be, reasons Newton. If the *effect* is the same – a downwards acceleration of 9.8 metres per second per second – the *cause* must be the same. And so: *the*

---

* The square of the period of revolution is proportional to the cube of the radius of the orbit. This law was shown to be correct not only for the planets orbiting the sun (Kepler), but also for the moons of Jupiter (Huygens). Newton assumes, by induction, that it should also hold for the hypothetical little moon orbiting the Earth. The constant of proportionality depends on the body around which the orbit is made: this is why data on the lunar orbit allow us to compute the period of the little moon.

† $a = v^2/r$, where v is the speed and r the radius of the orbit.

34

force which causes the little moon to turn around its orbit must be the same as that which causes objects to fall to the ground on Earth.

We call the force causing objects to fall gravity. Newton understands that it is this same gravity that makes the little moon turn around the Earth. Without this gravity it would run away in a straight line. But then, also, the real Moon must orbit the Earth because of gravity! And the moons which orbit Jupiter are attracted by Jupiter, and the planets which turn around the Sun are attracted by the Sun! Without this attraction, every celestial body would move in a straight line. So the universe, then, is a large space where bodies attract one another by means of forces; and there is a universal force: gravity – every body attracts every other body.

An immense vision takes shape. Suddenly, after millennia, there is no longer separation between the heavens and Earth; there is no 'natural level' for things, as Aristotle presumed; there is no centre to the world; things let free no longer seek their natural place but move in a straight line, for ever.

A simple calculation with the little moon allows Newton to deduce how the force of gravity changes with distance and to determine its strength,* given by what we call today Newton's constant, indicated by the letter 'G' for 'gravity'. On Earth, this force causes things to fall; in the heavens it holds planets and satellites on their orbits. The force is the same.

It's a subversion of the conceptual structure of the Aristotelian world, the dominant vision of the world throughout the Middle Ages. Think of Dante's universe, for example:

* $F = G \frac{M_1 M_2}{r^2}$.

just as for Aristotle, Earth is a ball at the centre of the universe, orbited by the celestial spheres. Not any more. The universe is an immense, infinite space studded with stars, without limit and without centre. Within it, material bodies run free and straight, unless a force, generated by another body, deviates them. Reference to ancient atomism is clear in Newton, even if still formulated in conventional terms:

> It seems probable to me that God, in the beginning, formed matter in solid, massy, hard, impenetrable, moveable particles, of such sizes and figures, and with such other properties, and in such proportions to space . . .[3]

The world of Newtonian mechanics is simple, and is summarized in figures 2.1 and 2.2. It is the world of Democritus reborn. A world made of a vast, undifferentiated space, always equal to itself, where particles run for ever and act upon each other – and nothing else. The world sung by Leopardi:

> . . . sitting here and gazing, I find that endless
> Spaces beyond that hedge, and more-than-human
> Silences, and the deepest peace and quiet
> Are fashioned in my thought . . .[4]

But now the vision is immensely more powerful than in Democritus, because it is not just a mental image with which

Figure 2.1 What is the world made of?

Figure 2.2 The world of Newton: particles which move
in space, as time passes, attracted by forces.

to order the world: it is now combined with mathematics,
with the legacy of Pythagoras, and with the mighty tradition
of the Alexandrian astronomers' mathematical physics. The
world of Newton is the world of Democritus, rendered
mathematical.

Newton acknowledges without hesitation the debt that
the new science owes to the science of antiquity. In the
first lines of his *The System of the World*, for example, he attrib-
utes to antiquity (correctly) the origins of the idea at the basis
of the Copernican revolution: 'It was the opinion of the
ancient philosophers that in the highest parts of the world
the stars remain fixed and motionless, and that the Earth
turns around the Sun'; although he is a bit confused as to
who did what in the past, and quotes – sometimes appositely,
sometimes out of context – Philolaus, Aristarchus of Samos,
Anaximander, Plato, Anaxagoras, Democritus and (!) 'the
learned Numa Pompilius, king of the Romans'.

The power of the new Newtonian intellectual frame-work
proves to be beyond all expectation. The entire technology
of the nineteenth century and of our own modern world
rests largely upon Newton's formulae. Three centuries have
passed, but it's still thanks to theories based upon Newton's

equations that today we build bridges, trains and skyscrapers, engines and hydraulic systems; that we know how to fly planes, make weather forecasts, predict the existence of a planet before seeing it and send spaceships to Mars . . . The modern world would not have been born without passing by way of Newton's little moon.

A new conception of the world, a way of thinking which raised the enthusiasm of the Enlightenment of Voltaire and Kant, an effective way for predicting the future: this has been and continues to be the immense legacy of the Newtonian revolution.

It seemed thus that the final key to understanding reality had been discovered: the world consists only of a great, infinite space where, as time passes, particles move and attract one another by means of forces. We can write precise equations that describe these forces. These equations prove to be immensely effective. Still in the nineteenth century, people said that Newton was not only one of the most intelligent and far-seeing of men but also the most fortunate – because there is only one system of fundamental laws, and he was given the good fortune of discovering it. Everything seemed clear.

But is that really all?

Michael: fields and light

Newton knew that his equations did not describe *all* the forces that exist in nature. There are forces other than gravity that act upon bodies. Things don't move just when they fall. The first problem left open by Newton was to understand the *other* forces that determine what happens around us. This had to wait until the nineteenth century, and it led to two surprises.

The first surprise is that almost all phenomena we see are governed by *a single* force, other than gravity: the force that today we call electromagnetism. It is this force which holds together the matter that forms solid bodies; holds together atoms in molecules, and electrons in atoms. This is what makes chemistry and living matter work. It is this force which operates in the neurons of our brain and governs our processing of the information on the world we perceive, and the way we think. And it's always this force that creates the friction which stops a sliding object, which softens the landing of a parachutist, which turns electric motors and combustion engines,* or that allows us to turn on lights and listen to the radio.

The second and biggest surprise, crucial to the story I'm telling, is that understanding this force requires an important modification to the world of Newton: the modification out of which modern physics was born, and the most important notion to keep in focus, to understand the rest of this book – the notion of field.

The understanding of how electromagnetic force works was made by another Briton, or rather by two: science's oddest couple – Michael Faraday and James Clerk Maxwell.

Michael Faraday is an impoverished Londoner without formal education, who works first in a bookbindery, then in a laboratory, where he excels, gains his master's confidence and grows into the most brilliant experimenter of nineteenth-century physics and its greatest visionary.

---

* The energy released by combustion engines is chemical and therefore, ultimately, electromagnetic.

Figure 2.3 Michael Faraday and James Clerk Maxwell.

Without knowing mathematics, he writes one of the best books of physics ever written, virtually devoid of equations. He sees physics with his mind's eye, and with his mind's eye creates worlds. James Clerk Maxwell is a rich Scottish aristocrat and one of the greatest mathematicians of the century. Despite being separated by a gulf in intellectual style as well as social origin, they succeed in understanding each other – and, together, combining two kinds of genius, they open the way to modern physics.

What was known about electricity and magnetism at the beginning of the eighteenth century consisted of little other than a few amusing sideshow tricks: glass rods which attract pieces of paper; magnets which repel and attract. The study of electricity and magnetism continued slowly throughout the eighteenth century and into the nineteenth, where we find Faraday working in a London laboratory full of bobbins, needles, knives and iron cages, exploring how electrical and magnetic things attract and repel. A good Newtonian, he tries to understand the force which acts between charged and

magnetic things. But slowly, guided by his hands, in close contact with these objects, he is led to an intuition that will become
the basis of modern physics. He 'sees' something new.

His intuition is this: we must not think of forces acting
directly between distant objects, as Newton presumed. We
must instead think that there exists an entity diffused throughout space, which is modified by electric and magnetic bodies
and which, in turn, acts upon (pushes and pulls) the bodies.
This entity, whose existence Faraday intuits, is today called
the field.

What is it, then, a field? Faraday sees it as formed by bun-
dles of very thin lines (infinitely thin), which fill space; an
invisible, gigantic cobweb filling everything around us. He
calls these lines 'lines of force' because, in some way, these
lines 'carry the force': they transmit the electric and the magnetic forces from one body to another, as if they were cables
pulling and pushing (figure 2.4).

An object with an electric charge (a rubbed glass rod, for
instance) distorts the electric and magnetic fields (the lines)

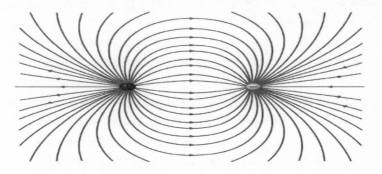

Figure 2.4 The field's lines fill space. Through them, two objects
with electrical charge interact. The force of the two objects
is 'carried' by the field's lines of force.

around itself, and in turn these fields produce a force on each charged object immersed in them. Thus, two distant charged objects do not attract or repel each other *directly* but only via the medium interposed between them.

If you take two magnets in your hands and play with them, bringing them together and apart repeatedly, feeling the force with which they attract and repel, it is not difficult to experience the same intuition as Faraday, to 'feel' via these effects the *field* interposed between the magnets.

This is a radically different idea from the Newtonian notion of a force acting between distant bodies. But it would have appealed to Newton. Newton, indeed, was perplexed regarding this very attraction at a distance that he himself had introduced. How does the Earth manage to attract the Moon, which is so distant? How can the Sun attract the Earth without coming into contact with it? He had written in a letter:

It is inconceivable that inanimate brute matter should, without the intervention of something else which is not material, operate upon and affect other matter, and have an effect upon it, without mutual contact.[5]

And, further down the page, we even find:

That Gravity should be innate, inherent and essential to Matter, so that one Body may act upon another at a Distance thro' a Vacuum, without the Mediation of any thing else, by and through which their Action and Force may be conveyed from one to another, is to me so great an Absurdity, that I believe no Man who has in philosophical Matters a competent Faculty of thinking, can ever fall into it. Gravity must be caused by an Agent acting constantly according to certain Laws; but whether this Agent be material or immaterial, I have left to the Consideration of my Readers.[6]

Newton is deeming his very own masterwork to be absurd – the very same work which was to be praised for centuries to come as the ultimate achievement of science! He understands that behind the action at a distance of his theory there must be something else, but he has no idea what, and leaves the question . . . *'to the Consideration of my Readers'*!

It is characteristic of genius to be aware of the limitations of its own findings, even in the case of such momentous outcomes as Newton's discovery of the laws of mechanics and universal gravity. Newton's theory worked so well, it turned out to be so useful, that for two centuries no one bothered any longer to question it – until Faraday, the 'reader' to whom Newton had bequeathed the unanswered question, found the key to understanding how bodies can attract and repel each other at a distance in a reasonable manner. Einstein will later apply Faraday's brilliant solution to Newton's own theory of gravity.

Introducing the new entity – the field – Faraday departs radically from Newton's elegant and simple ontology: the world is no longer made up only of particles that move in space while time passes. A new actor – the field – appears on the scene. Faraday is aware of the importance of the step he is taking. There are beautiful passages in his book where he asks whether these lines of force could be things with a real existence. After doubts and different considerations, he concludes that he thinks they are indeed real, but with 'the hesitation that is necessary when faced with the deepest

Figure 2.5 The world of Faraday and Maxwell: particles and fields which move in space, with the passage of time.

questions of science'.[7] He is conscious that he is suggesting nothing less than a modification of the structure of the world, after two centuries of uninterrupted successes for Newtonian physics (figure 2.5).

Maxwell quickly realizes that gold has been struck with this idea. He translates Faraday's insight, which Faraday explains only in words, into a page of equations.* These are now known as Maxwell's equations. They describe the behaviour of the electric and the magnetic fields: the mathematical version of the 'Faraday lines'.†

Today, Maxwell's equations are used daily to describe all electric and magnetic phenomena, to design antennae, radios, electric engines and computers. And this is not all: these same equations are needed to explain how atoms function (they are held together by electrical forces), and why the

* The equations fill a page in Maxwell's original treatise. Today the same equations can be written in half a line: $dF = 0$, $d*F = J$. We'll soon see why.

† If you visualize the field as a vector (an arrow) at each point of space, the point of the arrow indicates the direction of the Faraday lines, that is to say, the tangent of the Faraday lines, and the length of the arrow is proportional to the density of the Faraday lines.

particles of the material that forms a stone adhere together, or how the Sun works. They describe an amazing number and range of phenomena. Almost everything that we witness taking place – with the exception of gravity, but little else besides – is well described by Maxwell's equations.

But there is more. There is still what is perhaps the most beautiful success of science: Maxwell's equations tell us what light is.

Maxwell realizes that his equations predict that Faraday's lines can tremble and undulate, just like the waves of the sea. He computes the speed at which the undulations of Faraday's lines move and the result turns out to be . . . the same as for light! Why? Maxwell understands: because light is nothing other than this rapid trembling of Faraday's lines! Not only have Faraday and Maxwell figured out how electricity and magnetism work but, with the same stroke, as a collateral effect, they have figured out what light is.

We see the world around us in colour. What is colour? Put simply, it is the frequency (the speed of oscillation) of the electromagnetic wave that light is. If the wave vibrates more rapidly, the light is bluer. If it vibrates a little more slowly, the light is redder. Colour, as we perceive it, is the psychophysical reaction of the nerve signal generated by the receptors of our eyes, which distinguish electromagnetic waves of different frequencies.

I wonder how Maxwell felt when he realized that his equations – written to describe bobbins, small cages and little needles in Faraday's lab – turned out to explain the nature of light and colour . . .

Light is thus nothing more than a rapid vibration of the spiderweb of Faraday's lines, which ripple like the surface of a lake as the wind blows. It isn't true that we 'do not see'

Faraday lines. We *only* see vibrating Faraday lines. 'To see' is to perceive light, and light is the movement of Faraday lines. Nothing leaps from one location in space to another without something transporting it. If we see a child playing on the beach, it is only because between them and ourselves there is this lake of vibrating lines which transport their image to us. Is the world not marvellous?

The discovery is extraordinary, but there is even more. The ultimate corollary of the discovery has a concrete value for us that is unequalled. Maxwell recognizes that the equations foresee that Faraday's lines can also vibrate at much lower frequencies, that is to say, more slowly than light. Therefore, there must be *other* waves which nobody had yet seen, produced by the movement of electrical charges, which in turn move electrical charges. It must be possible to shake an electric charge *here*, and to produce a wave which will drive an electric current *there*. Only a few years later, these waves, anticipated theoretically by Maxwell, will be revealed by the German physicist Heinrich Hertz; and just a few years later still, Guglielmo Marconi builds the first radio.

All modern communications technology – radio, television, telephones, computers, satellites, wi-fi, the internet, etc. – is an application of Maxwell's prediction; the Maxwell equations are the basis for all calculations made by telecommunications engineers. The contemporary world, based on communications, emerges from the intuitions of a poor London bookbinder – a skilful explorer of ideas with a vivid imagination – who saw some lines in his mind's eye; and the work of a good mathematician who translated this vision into equations, understanding that in the blink of an eye the waves of these lines can carry news from one side of the planet to the other.

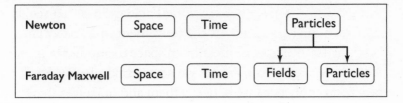

Figure 2.6 What is the world made of?

Our entire current technology is founded on the use of a physical thing – electromagnetic waves – which was not discovered empirically: it was predicted by Maxwell, simply by searching for the mathematical description accounting for the intuition Faraday got from bobbins and needles. This is the outstanding power of theoretical physics.

The world has changed: it is no longer made up of particles in space but of particles and fields in space (figure 2.6). It seems a minor change, but a few decades later a young Jewish man, a citizen of the world, will draw from it consequences that will go far beyond Michael Faraday's already fervid imagination, and which will shake Newton's world even further to its core.

PART TWO
The Beginning of the Revolution

Twentieth-century physics has radically modified the Newtonian image of the world. The new steps are the basis of a great deal of today's technology. The deepening of our understanding of the world is based on two theories: general relativity and quantum mechanics. Both demand a daring re-evaluation of our conventional ideas about the world: space and time in relativity; matter and energy in quantum theory.

In this part of the book I describe the two theories in some detail, trying to clarify their core meaning and highlight the conceptual revolution they brought about. It's here that the magic of twentieth-century physics begins. Studying and trying to understand them in depth is a bewitching adventure.

These two theories — relativity and the quanta — provide the basis on which we are today building a quantum theory of gravity. They are the ground from which we are trying to move forwards.

3. Albert

The father of Albert Einstein built power stations in Italy. When Albert was a young boy, the Maxwell equations were only a few decades old – but Italy was entering its industrial revolution, and the turbines and transformers that his father constructed were already based on them. The power of the new physics was obvious.

Albert was a rebel. His parents left him in Germany to attend high school, but he found the German school system too rigid and militaristic; he could not stand the authority of the school and abandoned his studies. He joined his parents in Italy, in Pavia, and spent his time loafing. Later, he went to study in Switzerland, initially failing to get into the Zurich Polytechnic, as he wished. After his university years he could not find a research position and, in order to live with the girl he loved, he found employment in the patent office of Berne.

It wasn't much of a profession for a physics graduate, but it gave Albert time to think, and to work independently. And he did think and work. After all, this is what he had done since his early youth: he would read Euclid's *Elements* and Kant's *Critique of Pure Reason* instead of attending to what he was being taught at school. You don't get to new places by following established tracks.

At the age of twenty-five, Einstein sends to the *Annalen der Physik* three articles. Each was worthy of a Nobel Prize, and more. Each one of the three is a pillar supporting our understanding of the world. I have already spoken of the first article, in which the young Albert calculates the dimensions of atoms and proves, after twenty-three centuries, that the ideas of Democritus were correct: matter is granular.

The second article is the one for which Einstein is most famous – the article in which he introduces the theory of relativity – and it is to the theory of relativity that this chapter is devoted.

In fact, there are two theories of relativity. The envelope sent by the twenty-five-year-old Einstein contained the exposition of the first of these: the theory known today as 'special relativity'. This is an important clarification of the structure of space and time, which I illustrate here before turning to the other, and most important of Einstein's theories: general relativity.

Special relativity is a subtle and conceptually difficult theory. It is more difficult to digest than general relativity. Reader, don't become demoralized if the next few pages sound a bit abstruse. The theory shows, for the first time, that in the Newtonian vision of the world there isn't just something missing: rather, it must be radically modified – in a way that goes completely against common sense. It is the first real leap into the revision of our most intuitive understanding of the world.

The extended present

The theories of Newton and of Maxwell appear to contradict each other in a subtle way. Maxwell's equations determine a velocity: the velocity of light. But Newton's mechanics is not compatible with the existence of a fundamental velocity, because what enters Newton's equations is acceleration, not velocity. In Newton's physics, velocity can only be velocity of something *with respect to something else*. Galileo had underlined the fact that the Earth moves with respect to the Sun, even if we do not perceive this movement, because what we usually term 'velocity' is velocity 'with respect to Earth'. Velocity, we say, is a *relative* concept, that is, there is no meaning to the velocity of an object by itself: the only velocity which exists is the velocity of an object with respect to another object. This is what physics students learned in the nineteenth century, and what they learn today. But if this is so, then the speed of light determined by Maxwell's equations is velocity with respect to what?

One possibility is that there is a kind of universal substratum in relation to which light moves and has its speed. But the predictions of Maxwell's theory seem to be independent of this substratum. The experimental attempts to measure the speed of the Earth with respect to this hypothetical substratum tried at the end of the twentieth century all failed.

Einstein has claimed that he was not put on the right track by any experiments but only by reflecting on the apparent contradiction between Maxwell's equations and Newton's mechanics. He asked himself whether there was a way of rendering Newton's and Galileo's core discoveries and Maxwell's theory consistent.

In doing so, Einstein arrives at a stupefying discovery. To understand it, think of all the past, present and future events (with respect to the moment in which you are reading) and imagine them distributed as in figure 3.1.

Well, Einstein's discovery is that this diagram is incorrect. In reality, things are actually as they are depicted in figure 3.2.

Between the past and the future of an event (for example, between the past and the future for you, where you are, and

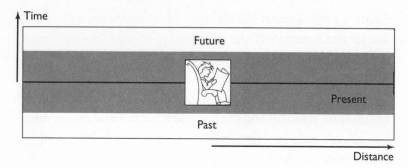

Figure 3.1 Space and time before Einstein.

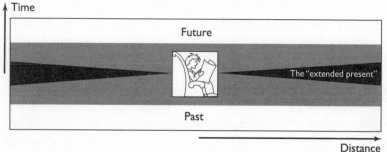

Figure 3.2 The structure of 'spacetime'. For every observer,
the 'extended present' is the intermediate zone between
the past and the future.

in the precise moment in which you are reading) there exists
an 'intermediate zone', an 'expanded present'; a zone that is
neither past nor future. This is the discovery made with special relativity.

The duration of this intermediate zone,* which is neither
in your past nor in your future, is very small and depends on
where an event takes place relative to you, as illustrated in

* The set of events at a space-like distance from a reference event.

figure 3.2: the greater the distance of the event from you, the longer the duration of the extended present. At a distance of a few metres from your nose, dear reader, the duration of what for you is the intermediate zone, neither past nor future, is no more than a few nanoseconds: next to nothing (the number of nanoseconds in a second is the same as the number of seconds in thirty years). This is much less than we could possibly notice. On the other side of the ocean, the duration of this intermediate zone is a thousandth of a second, still well below the threshold of our perception of time – the minimum amount of time we perceive with our senses – which is somewhere in the order of a tenth of a second. But on the Moon the duration of the expanded present is a few seconds, and on Mars it is a quarter of an hour. This means we can say that, on Mars, there are events that in this precise moment have already happened, events that are yet to happen, but also a quarter of an hour during which things occur that are neither in our past nor in our future.

They are elsewhere. We had never before been aware of this 'elsewhere' because, next to us, this 'elsewhere' is too brief; we are not quick enough to notice it. But it exists, and it is real.

This is why it is impossible to hold a smooth conversation between here and Mars. Say I am on Mars and you are here. I ask you a question and you reply as soon as you've heard what I said; your reply reaches me a quarter of an hour after I posed the question. This quarter of an hour is time that is neither past nor future to the moment in which you've replied to me. The key fact about nature that Einstein understood is that this quarter of an hour is inevitable: there is no way of reducing it. It is woven into the texture of the events of

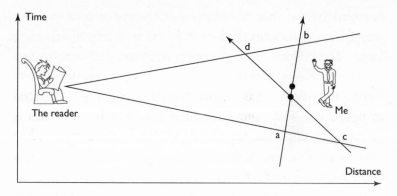

Figure 3.3 The relativity of simultaneity.

space and of time: we cannot abbreviate it, any more than we can send a letter to the past.

It's strange, but this is how the world happens to be. As strange as the fact that in Sydney people live upside down: strange, but true. One gets accustomed to the fact, which then becomes normal and reasonable. It is the structure of space and time that is made like this.

This implies that it makes no sense to say of an event on Mars that it is taking place 'just now', because 'just now' does not exist (figure 3.3).* In technical terms, we say that Einstein

---

* The astute reader will object that the *halfway* moment of my quarter of an hour can be considered simultaneous to your reply. The reader who has studied physics will recognize that this is 'Einstein's convention' for defining simultaneity. This definition of simultaneity depends on how I move, and consequently does not define simultaneity between two events but only a simultaneity *relative* to the state of movement of particular bodies. In figure 3.3 a dot is halfway between *a* and *b*, the points at which I exit from the past of the observer and enter his future. The other dot is halfway between *e* and *d*, the points at which I exit from the past of the observer and enter into his future if I move along a different

has understood that 'absolute simultaneity' does not exist: there is no collection of events in the universe which exist 'now'. The collection of all the events in the universe cannot be described as a succession of 'now's, of presents, one following the other; it has a more complex structure, illustrated in figure 3.2. The figure describes that which in physics is called spacetime: the set of all past and future events, but also those that are 'neither-past-nor-future'; these do not form a single instant: they have themselves a duration.

In the Andromeda Galaxy, the duration of this expanded present is (with respect to us) 2 million years. Everything that happens during these 2 million years is neither past nor future with respect to ourselves. If a friendly advanced Andromeda civilization decided to send a fleet of spacecraft to visit us, it would make no sense to ask whether 'now' the fleet has already left, or not yet. The only meaningful question is when we receive the first signal from the fleet: from that moment on – not earlier – the departure of the fleet is in our past.

The discovery of the structure of spacetime made by the young Einstein in 1905 has concrete consequences. The fact that space and time are intimately connected, as in figure 3.2, implies a subtle restructuring of Newton's mechanics, which Einstein rapidly completes in 1905 and 1906. A first result of this restructuring is that, as space and time fuse together in a single concept of spacetime, so the electric field and the magnetic fields fuse together in the same way, merging into a single entity which today we call the electromagnetic field.

---

trajectory. Both dots are simultaneous as regards the reader, according to this definition of 'simultaneity', but they occur in successive times. The two dots are each simultaneous to the reader, but relative to two different motions of mine. Hence the term 'relativity'.

The complicated equations written by Maxwell for the two fields become simple when written in this new language.

There is another implication of the theory, freighted with heavy consequences. The concepts of 'energy' and 'mass' become combined in the same way as time and space, and electric and magnetic fields, are fused together in the new mechanics. Before 1905 two general principles appeared certain: conservation of mass, and conservation of energy. The first had been extensively verified by chemists: mass never changes in a chemical reaction. The second – conservation of energy – followed directly from Newton's equations and was considered one of the most *incontrovertible* laws. But Einstein realizes that energy and mass are two facets of the same entity, just as the electric and magnetic fields are two facets of the same field, and as space and time are two facets of the one thing: spacetime. This implies that mass, by itself, is not conserved; and energy – as it was conceived at the time – is not independently conserved either. One may be transformed into the other: only one single law of conservation exists, not two. What is conserved is the sum of mass and energy, not each separately. Processes must exist that transform energy into mass, or mass into energy.

A rapid calculation teaches Einstein how much energy is obtained by transforming one gram of mass. The result is the celebrated formula $E = mc^2$. Since the speed of light c is a very large number, and c^2 an even greater number, the energy obtained from transforming one gram of mass is enormous; it is the energy of millions of bombs exploding at the same time – enough energy to illuminate a city and power the industries of a country for months or, conversely, capable of destroying in a second hundreds of thousands of human beings, in a city such as Hiroshima.

The theoretical speculations of the young Einstein had transported humanity into a new era: the era of nuclear power, an era of new possibilities, and new dangers. Today, thanks to the intelligence of a rebellious young man who would not abide rules, we have the instruments to bring light to the homes of the 10 billion human beings who will soon inhabit the planet, to travel in space towards other stars, or to destroy each other and devastate the planet. It depends on our choices; on which leaders we call upon to decide for us.

Today the structure of spacetime proposed by Einstein is well understood and repeatedly tested in laboratories; it is considered conclusively established. Time and space are different from the way they had been conceived since Newton. Space does not exist independently from time. In the expanded space of figure 3.2 there is no particular slice having a better claim than others to be called 'space now'. Our intuitive idea of the present – the ensemble of all events happening 'now' in the universe – is an effect of our blindness: our inability to recognize small temporal intervals. It is an illegitimate extrapolation from our parochial experience.

The present is like the flatness of the Earth: an illusion. We imagined a flat Earth because of the limitations of our senses, because we cannot see much beyond our own noses. Had we lived on an asteroid of a few kilometres in diameter, like the Little Prince, we would have easily realized we were on a sphere. Had our brain and our senses been more precise, had we easily perceived time in nanoseconds, we would never have made up the idea of a 'present' extending everywhere. We would have easily recognized the existence of the intermediate zone between past and future. We would have realized that saying 'here and now' makes sense, but that saying 'now' to designate events 'happening now' throughout the universe makes no sense. It is like asking whether our galaxy is 'above or below' the galaxy of Andromeda: a question that makes no sense, because 'above' or 'below' has meaning on the surface of the Earth, not in the universe. There isn't an 'up' or a 'down' in the universe. Similarly, there isn't either always a 'before' and an 'after' between two events in the universe. The resulting knitted structure that space and time form together, depicted in figures like 3.2 and 3.3, is what physicists call 'spacetime' (figure 3.4).

When the *Annalen der Physik* published the article by Einstein in which all this was suddenly clarified, the impact upon the world of physics was momentous. The apparent contradiction between the equations of Maxwell and Newtonian physics were well known, and no one knew how to resolve them. Einstein's solution, astonishing and extremely elegant, took everyone by surprise. The story goes that in the dimly lit old halls of Cracow University, an austere professor of physics came out of his study, waving around Einstein's article, screaming, 'The new Archimedes is born!'

But despite the outcry provoked by the step forwards made by Einstein in 1905, we are not yet at his masterpiece.

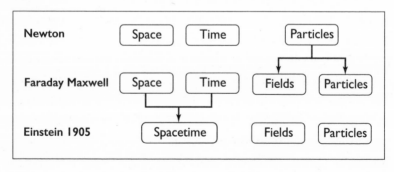

Figure 3.4 What is the world made of?

Einstein's triumph is the *second* theory of relativity, the theory of *general relativity*, published ten years later, when he was thirty-five.

The theory of 'general relativity' is the most beautiful theory produced by physics, and the first of the pillars of quantum gravity. It is at the heart of the narrative of this book. Here, the real magic of twentieth-century physics begins.

The most beautiful of theories

After publishing the theory of special relativity, Einstein becomes a renowned physicist and receives offers of work from numerous universities. But something troubles him: special relativity does not square with what was known about gravity. He realizes this while writing a review on his theory, and wonders whether the venerable theory of the 'universal gravity' of the father of physics, Newton, should not be reconsidered as well, to make it compatible with his relativity.

The origin of the problem is easy to understand. Newton had tried to explain why things fall and planets revolve. He had imagined a 'force' that draws all bodies towards one another: the 'force of gravity'. How this force managed to draw distant things together without anything between them was not understood. Newton himself, as we have seen, had suspected that in the idea of a force acting between distant bodies that do not touch there was something missing; and that in order for the Earth to attract the Moon something that could transmit this force had to be there between the two. Two hundred years later, Faraday had found the solution – not for the force of gravity, but for the electric and magnetic forces: the field. Electric and magnetic fields 'carry around' the electric and magnetic force.

It's clear, at this stage, to any reasonable person, that the force of gravity must have its Faraday lines as well. It's clear also, by analogy, that the force of attraction between the Sun and the Earth, or between the Earth and falling objects, must be attributed to a field – in this case, a gravitational field. The solution discovered by Faraday and Maxwell to the question as to what carries the force must reasonably be applied not only to electricity but also to gravity. There must be a gravitational field, and some equations analogous to Maxwell's, capable of describing how Faraday's gravitational lines move. In the first years of the twentieth century this is clear to any sufficiently reasonable person; that is to say, only to Albert Einstein.

Einstein, fascinated since adolescence by the electromagnetic field that pushed the rotors in his father's power stations, begins to look into this gravitational field and search for what kind of maths could describe it. He immerses himself in the problem. It would take ten years to resolve. Ten years of manic studies, attempts, mistakes, confusion, brilliant ideas, wrong ideas, a long series of articles published with incorrect equations, further mistakes and stress. Finally, in 1915, he commits to print an article containing the complete solution, which he names the General Theory of Relativity: his masterpiece. It is Lev Landau, the most outstanding theoretical physicist of the Soviet Union, who called it 'the most beautiful of theories'.

The reason for the beauty of the theory is not hard to see. Instead of simply inventing the mathematical form of the gravitational field and seeking to devise the equations for it, Einstein fishes out the other unresolved question in the furthest depths of Newton's theory and combines the two questions.

Newton had returned to Democritus's idea, according to which bodies move in *space*. This *space* had to be a large, empty container, a rigid box for the universe; an immense scaffolding in which objects run in straight lines, until a force causes them to curve. But what is this 'space' which contains the world made of? What *is* space?

To us, the idea of space seems natural, but it is our familiarity with Newtonian physics that makes it so. If you think about it, empty space is not part of our experience. From Aristotle to Descartes, that is to say, for two millennia, the Democritean idea of space as a peculiar entity, distinct from things, had never been seen as reasonable. For Aristotle, as for Descartes, things have extension: extension is a property of things; extension does not exist without something being extended. I can take away the water from a glass, but air will fill it. Have you ever seen a really empty glass?

If between two things there is *nothing*, Aristotle reasoned, then there is nothing. How can there be at the same time something (space) and nothing? What is this empty space within which particles move? Is it something, or is it nothing? If it is nothing, it doesn't exist, and we can do without it. If it is something, can it be true that its only property is to be there, *doing* nothing?

Since antiquity, the idea of empty space, halfway between a thing and a non-thing, had troubled thinkers. Democritus himself, who had placed empty space at the basis of his world where atoms course, certainly wasn't crystal clear on the issue: he wrote that empty space is something 'between being and non-being': 'Democritus postulated the full and the empty, calling one "Being", and the other "Non-Being",' says Simplicius.[1] Atoms are being. Space is non-being – a 'non-being' that, nevertheless, exists. It is difficult to be more obscure than this.

Newton, who resuscitated the Democritean idea of space, had tried to patch things up by arguing that space was God's *sensorium*. No one has ever understood what Newton meant by 'God's sensorium', perhaps not even Newton himself. Certainly, Einstein, who gave little credit to the idea of a God (with or without a sensorium), except as a playful rhetorical device, found Newton's explanation of the nature of space utterly unconvincing.

Newton struggled considerably to overcome the scientists' and philosophers' resistance to his reviving the Democritean concept of space; at first nobody took him seriously. Only the extraordinary efficacy of his equations, which turned out to predict always the correct outcome, ended up silencing criticism. But doubts concerning the plausibility of the Newtonian concept of space persisted, and Einstein, who read philosophers, was well aware of them. Ernst Mach, whose influence Einstein readily acknowledged, was the philosopher who highlighted the conceptual difficulties of the Newtonian idea of space – the same Mach who did not believe in the existence of atoms. (A good example, incidentally, of how the same person can be short-sighted in one respect and far-seeing in another.)

Thus, Einstein addresses not one but two problems. First, how can we describe the gravitational field? Second, what is Newton's space?

And it's here that Einstein's extraordinary stroke of genius occurs, one of the greatest flights in the history of human thinking: what if the gravitational field turned out actually to *be* Newton's mysterious space? What if Newton's space

64

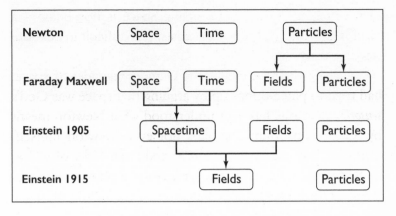

Figure 3.5 What is the world made of?

was nothing more than the gravitational field? This extremely simple, beautiful, brilliant idea is the theory of general relativity.

The world is not made up of space + particles + electromagnetic field + gravitational field. The world is made up of particles + fields, and nothing else; there is no need to add space as an extra ingredient. Newton's space *is* the gravitational field. Or vice versa, which amounts to saying the same thing: the gravitational field is space (figure 3.5).

But, unlike Newton's space, which is flat and fixed, the gravitational field, by virtue of being a field, is something which moves and undulates, subject to equations – like Maxwell's field, like Faraday's lines.

It is a momentous simplification of the world. Space is no longer different from matter. It is one of the 'material' components of the world, akin to the electromagnetic field. It is a real entity which undulates, fluctuates, bends and contorts.

We are not contained within an invisible, rigid scaffolding:

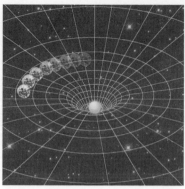

Figure 3.6 The Earth turns around the Sun because
spacetime around the Sun is curved, rather like a bead
which rolls on the curved wall of a funnel.

we are immersed in a gigantic, flexible mollusc (the metaphor is Einstein's). The Sun bends space around itself, and
the Earth does not circle around it drawn by a mysterious
distant force but runs straight in a space that inclines. It's like
a bead which rolls in a funnel: there are no mysterious forces
generated by the centre of the funnel, it is the curved nature
of the funnel wall which guides the rotation of the bead.
Planets circle around the Sun, and things fall, because space
around them is curved (figure 3.6).

A little more precisely, what curves is not space but
spacetime – that spacetime which, ten years previously, Einstein himself had shown to be a structured whole rather than
a succession of instants.

This is the idea. Einstein's only problem was to find the
equations to make it concrete. How to describe this bending
of spacetime? And here Einstein is lucky: the problem had
already been solved by the mathematicians.

The greatest mathematician of the nineteenth century,

Figure 3.7 A curved (bidimensional) surface.

Carl Friedrich Gauss, the 'prince of mathematicians', had written maths to describe curved surfaces, such as the surfaces of hills, or such as the one portrayed in figure 3.7.

Then he had asked a talented student of his to generalize this maths to curved spaces in three or more dimensions. The student, Bernhard Riemann, produced a ponderous doctoral thesis of the kind that seems completely useless.

Riemann's result was that the properties of a curved space (or spacetime) in any dimension are described by a particular mathematical object, which we now call Riemann curvature and indicate with the letter 'R'. If you think of a landscape of plains, hills and mountains, the curvature R of the surface is zero in the plains, which are flat – 'without curvature' – and different from zero where there are valleys and hills; it is at its maximum where there are pointed peaks of mountains, that is to say, where the ground is least flat, or most curved. Using Riemann's theory, it is possible to describe the shape of curved spaces in three or four dimensions.

With a great deal of effort, seeking help from friends better versed in mathematics than himself, Einstein learns Riemann's maths – and writes an equation where R is proportional to the energy of matter. In words: spacetime curves more where there is matter. That is it. The equation is the analogue of the Maxwell equations, but for gravity rather than electricity. The equation fits into half a line, and there is nothing more. A vision – that space curves – becomes an equation.

But within this equation there is a teeming universe. And here the magical richness of the theory opens up into a phantasmagorical succession of predictions that resemble the delirious ravings of a madman but which have all turned out to be true. Even up to the beginning of the 1980s, almost nobody took the majority of these fantastical predictions entirely seriously. And yet, one after another, they have all been verified by experience. Let's consider a few of them.

To begin with, Einstein recalculates the effect of a mass like the Sun on the curvature of the space that surrounds it, and the effect of this curvature on the movements of the planets. He finds the movements of the planets as predicted by Kepler's and Newton's equations, but not exactly: in the vicinity of the Sun, the effect of the curvature of space is stronger than the effect of Newton's force. Einstein computes the movement of Mercury, the planet closest to the Sun and hence the one for which the discrepancy between the predictions of his and Newton's theories is greatest. He finds a difference: the point of the orbit of Mercury closest to the Sun moves every year 0.43 seconds of arc more than that predicted by Newton's theory. It is a small difference, but, within the scope of what astronomers were able to measure, and comparing the predictions with the observations of astronomers, the verdict is unequivocal: Mercury follows the trajectory predicted by Einstein, not the one predicted by Newton. Mercury, the fleet-footed messenger of the gods, the god of the winged sandals, follows Einstein, not Newton.

Einstein's equation, then, describes how space curves very close to a star. Due to this curvature, light deviates. Einstein predicts that the Sun causes light to curve around it. In 1919 the measurement is achieved; a deviation of light is measured which turns out to be exactly in accordance with the prediction.

But it is not only space that curves: time does, too. Einstein predicts that time on Earth passes more quickly at higher altitude, and more slowly at lower altitude. This is measured, and also proves to be the case. Today we have extremely precise clocks, in many laboratories, and it is possible to measure this strange effect even for a difference in altitude of just a few centimetres. Place a watch on the floor and another on a table: the one on the floor registers less passing of time than the one on the table. Why? Because time is not universal and fixed, it is something which expands and shrinks, according to the vicinity of masses: the Earth, like all masses, distorts spacetime, slowing time down in its vicinity. Only slightly – but two twins who have lived respectively at sea-level and in the mountains will find that, when they meet up again, one will have aged more than the other (figure 3.8).

This effect offers an interesting explanation as to why things fall. If you look at a map of the world and the route taken by an aeroplane flying from Rome to New York, it does not seem to be straight: the aeroplane makes an arc towards the north. Why? Because, the Earth being curved, crossing northwards is shorter than keeping to the same parallel. The distances between meridians are shorter the more northerly you are; therefore, it is better to head northwards, to shorten the route (figure 3.9).

Well, believe it or not, a ball thrown upwards falls downwards for the same reason: it 'gains time' moving higher up,

Figure 3.8 Two twins spend their time one at sea-level and the other in the mountains. When they meet up again, the twin who lived in the mountains is older. This is the gravitational dilation of time.

because time passes at a different speed up there. In both cases, aeroplane and ball follow a straight trajectory in a space (or spacetime) that is curved (figure 3.10).*

But the predictions of the theory go well beyond these minute effects. Stars burn as long as they have available hydrogen – their fuel – then die out. The remaining material is no longer supported by the pressure of the heat and collapses under its own weight. When this happens to a large enough star, the weight is so strong that matter is squashed

* Airplane and ball follow a geodesic in a curved space. In the case of the ball, the geometry is approximately given by the metric $ds^2 = (1 - 2\Phi(x))\,dt^2 - dx^2$, where $\Phi(x)$ is the Newtonian potential. The effect of the gravitational field is reduced to the dilation of time with altitude. (The reader familiar with the theory will notice the curious sign inversion: the physical trajectory maximizes proper time.)

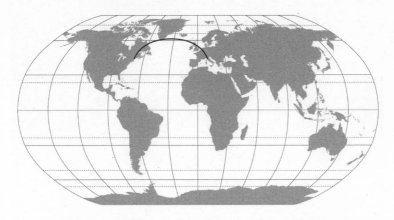

Figure 3.9 The further north you go, the smaller the distance between two meridians.

Figure 3.10 The higher up something is, the more quickly time passes for it.

down to an enormous degree and space curves so intensely as to plunge down into an actual hole. A black hole.

When I was a university student, black holes were regarded as a scarcely credible implication of an esoteric theory. Today they are observed in their hundreds and studied in detail by astronomers. One of these black holes, with a mass a million times greater than the Sun, is located at the centre of our galaxy – we can observe stars orbiting around it. Some, passing too close, are destroyed by its violent gravity.

Further still, the theory predicts that space ripples like the surface of the sea, and that these ripples are waves similar to the electromagnetic ones which make television possible. The effects of these 'gravitational waves' can be observed in the sky on binary stars: they radiate such waves, losing energy and slowly falling towards each other.* Gravitational waves produced by two black holes falling into one another were directly observed by an antenna on Earth in late 2015, and the announcement, given in early 2016, has once again left the world speechless. Once more, the seemingly mad predictions of Einstein's theory turn out to be precisely true.

And further still, the theory predicts that the universe is expanding and emerged from a cosmic explosion 14 billion years ago – a subject I will discuss in more detail shortly.

This rich and complex range of phenomena – bending of rays of light, modification of Newton's force, slowing down of clocks, black holes, gravitational waves, expansion of the universe, the Big Bang – follow from understanding that space is not a dull, fixed container but possesses its own dynamic, its own 'physics', just like the matter and the other fields it contains. Democritus himself would have smiled with pleasure, had he been able to see that his idea of space would turn out to have such an impressive future. It is true

---

* Observations of the binary system PSR B193+16 show that the two stars which revolve around one another radiate gravitational waves. These observations brought a Nobel Prize for Russell Hulse and Joseph Taylor in 1993.

that he termed it non-being, but what he meant by being (δέν) was matter; and he wrote that his non-being, the void, nevertheless 'has a certain physics (φύσιν) and a substantiality of its own'.* How right he was.

Without the notion of fields introduced by Faraday, without the spectacular power of mathematics, without the geometry of Gauss and Riemann, this 'certain physics' would have remained incomprehensible. Empowered by new conceptual tools and by mathematics, Einstein writes the equations which describe Democritus's void and finds for its 'certain physics' a colourful and amazing world where universes explode, space collapses into bottomless holes, time slows down in the vicinity of a planet, and the boundless expanses of interstellar space ripple and sway like the surface of the sea . . .

All of this sounds like a tale told by an idiot, full of sound and fury, signifying nothing. And yet, instead, it is a glance towards reality. Or better, a glimpse of reality, a little less veiled than our blurred and banal everyday view of it. A reality which seems to be made of the same stuff our dreams are made of, but which is nevertheless more real than our clouded daily dreaming.

And all this is the result only of an elementary intuition – that spacetime and the gravitational field are one and the same thing – and a simple equation which I can't resist copying out here, even if most of my readers will certainly not be able to decipher it. I do so, anyway, in the hope that they might be able to catch a glimpse of its beautiful simplicity:

$$R_{ab} - \frac{1}{2} R g_{ab} + \Lambda g_{ab} = 8 \pi G \, T_{ab}$$

* Plutarch, *Adversus colotem*, 4, 1108. The word φύσιν means 'nature', and includes the sense 'the nature of something'.

In 1915 the equation was simpler still, because the term $+\Lambda g_{ab}$, which Einstein added two years later (and which I discuss below) did not yet exist.* R_{ab} depends on Riemann's curvature, and together with $\frac{1}{2}Rg_{ab}$ represents the curvature of spacetime; T_{ab} stands for the energy of matter; G is the same constant that Newton found: the constant that determines the strength of the force of gravity.

That's it. A vision and an equation.

Mathematics or physics?

I would like to pause, before continuing with physics, to make a few observations about mathematics. Einstein was no great mathematician. He struggled with maths. He says this himself. In 1943 he replied in the following way to a nine-year-old child with the name of Barbara who wrote to him about her difficulties with the subject: 'Don't worry about experiencing difficulties with maths, I can assure you that my own problems are even more serious!'[2] It seems like a joke, but Einstein was not kidding. With mathematics, he needed help: he had it explained to him by patient fellow students and friends, such as Marcel Grossman. It was his intuition as a physicist that was prodigious.

During the last year in which he was completing the construction of his theory, Einstein found himself competing

---

* This term is called 'cosmological' because its effects occur only at an extremely large, or 'cosmological' distances. The constant Λ is called the 'cosmological constant', and its value was measured at the end of the 1990s, bringing a Nobel Prize in 2011 for the astronomers Saul Perlmutter, Brian P. Schmidt and Adam G. Riess.

with David Hilbert, one of the greatest mathematicians of all time. Einstein had given a lecture, attended by Hilbert, in Göttingen. Hilbert immediately understood that Einstein was in the process of making a major discovery, grasped the idea and tried to overtake Einstein and be the first to write the correct equations of the new theory Einstein was slowly building. The sprint to the finish line between the two giants was a nail-biting affair, eventually decided by a matter of just a few days. Einstein, in Berlin, ended up giving a public lecture almost every week, each time presenting a different equation, anxious that Hilbert would not get to the solution before him. The equation was incorrect every time. Until, that is, by a hair's breadth – just marginally ahead of Hilbert – Einstein found the right one. He had won the race.

Hilbert, a gentleman, never questioned Einstein's victory, even though he was working on very similar equations at the time. In fact, he left a gentle and beautiful phrase which captures perfectly Einstein's difficult relationship with mathematics, and, perhaps, the difficult relationship which exists generally between the whole of physics and mathematics. The maths that was necessary to formulate the theory was geometry in four dimensions, and Hilbert writes:

> Any youngster on the streets of Göttingen* understands geometry in four dimensions better than Einstein. And yet, it was Einstein who completed the task.

Why? Because Einstein had a unique capacity to *imagine* how the world might be constructed, to 'see' it in his mind.

---

* Göttingen, where Hilbert worked, was at this time the seat of the most important school of geometry.

The equations, for him, came afterwards; they were the language with which to make concrete his visions of reality. For Einstein, the theory of general relativity is not a collection of equations: it is a mental image of the world arduously translated into equations.

The idea behind the theory is that spacetime curves. If spacetime had only two dimensions, and we lived on a sort of plane, it would be easy to imagine what it means to say that 'physical space curves'. It would mean that the physical space in which we live is not like a flat table but resembles instead a surface with mountains and valleys. But the world we inhabit does not have only two dimensions, it has three. Four, in fact, when time is included. To imagine a curved space in four dimensions is more complicated, because in our habitual perception we do not have an intuition of a 'larger space' within which spacetime can curve. But Einstein's imagination had no difficulty in intuiting the cosmic mollusc in which we are immersed, which can be squashed, stretched and twisted – and that constitutes the spacetime around us. It is thanks to this visionary clarity that Einstein managed to be the first to write the theory.

In the end, a degree of tension between Hilbert and Einstein did develop. A few days before Einstein made his successful equation public, Hilbert had sent an article to a periodical which shows just how close he had come to the same solution – and even today historians of science are faced with doubts when trying to evaluate the respective contributions of these two giants. At some point their relations cooled, and Einstein feared that Hilbert, more senior and powerful than him, would seek to attribute to himself too much of the merit for the construction of the theory. But Hilbert never claimed to be the first to discover general relativity – and in a world such as that of science, where often, too frequently, disputes over precedence become poisonous – the two gave a truly wonderful example of wisdom, clearing the field of all negative tension.

Einstein writes a marvellous letter to Hilbert, summarizing the profound sense of the shared course they had taken:

> There was a moment in which something like an irritation came between us, the origin of which I no longer want to analyse. I have fought against the bitterness which it provoked in me, and have succeeded completely in doing so. I again think of you with unclouded friendship, and I ask you to do the same for me. It is really a pity if companions such as we are, who have managed to forge a path aside from the pettiness of this world, could find anything other than joy in each other's company.[3]

The cosmos

Two years after the publication of his equation, Einstein decides to use it to describe the space of the entire universe, considered at the largest scale. And here he has another of his amazing ideas.

For thousands of years, men had asked themselves whether the universe was infinite, or had a limit. Both hypotheses entail thorny problems. An infinite universe does not seem to stand to reason: if it is infinite, for example, there must exist somewhere a reader just like you who is reading the very same book (infinity is truly vast, and there are not sufficient combinations of atoms to fill it with things always different from each other). In fact, there must be not only one but an infinite series of readers identical to yourself . . . But if there is a limit to the universe, what is that boundary? What sense is there in a border with nothing on the other side? Already in the sixth century CE, in Taranto, the Pythagorean philosopher Archytas had written:

> If I found myself in the furthest sky, that of the fixed stars, would I be able to stretch my hand, or a rod, out beyond it – or not? That I should not be able to is absurd; but if I am able to, then an outside exists, be it of matter, or space. In this way one could proceed ever further, towards the end, from time to time asking the same question, as to whether there will always be something into which to extend the rod.[4]

These two absurd alternatives – the absurdity of an infinite space, and the absurdity of a universe with a fixed border – didn't seem to leave any reasonable choice between them.

But Einstein finds a third way: the universe can be *finite* and at the same time *have no boundary*. How? Just as the surface of the Earth is not infinite but does not have a boundary either, where it 'ends'. This can happen, naturally enough, if something is curved: the surface of the Earth is curved. And in the theory of general relativity, of course, three-dimensional space can also be curved. Consequently, our universe can be finite but borderless.

On the surface of the Earth, if I were to keep walking in a straight line, I would not advance ad infinitum: I would eventually get back to the point I started from. Our universe could be made in the same way: if I leave in a spacecraft and journey always in the same direction, I fly around the universe and eventually end up back on Earth. A three-dimensional space of this kind, finite but without boundary, is called a 3-sphere.

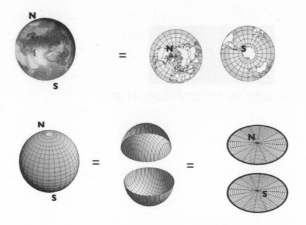

Figure 3.11 A sphere can be represented as two discs which in reality are smoothly joined all along their edges.

To understand the geometry of a '3-sphere', let us return to the ordinary sphere; the surface of a ball, or the Earth. To represent the surface of the Earth on a plane, we can draw two discs, as is customary when drawing the continents (figure 3.11).

Notice that an inhabitant of the southern hemisphere is in a certain sense 'surrounded' by the northern hemisphere, since in whichever direction she goes to exit her hemisphere, she will always arrive in the other one. But the contrary is obviously true as well: each hemisphere 'surrounds', and is surrounded by, the other. A 3-sphere may be represented in a similar fashion, but with everything given an additional dimension: two balls stuck together all along their edges (figure 3.12).

When we leave one ball we enter into the other, just as when we leave one of the discs in the representation of the globe we enter into the other disc. Each ball surrounds and is surrounded by the other. Einstein's idea is that space could be a 3-sphere: something with a finite volume (the sum of the

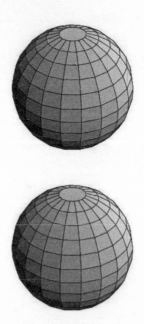

Figure 3.12 A 3-sphere can be represented as
two balls joined together.

volume of the two balls), but without borders.* The 3-sphere
is the solution which Einstein proposes in his work of 1917
to the problem of the border of the universe. This article
initiates modern cosmology, the study of the entire visible
universe, studied at the grandest scale. From it will arise
the discovery of the expansion of the universe; the theory of
the Big Bang; the problem of the birth of the universe, and
much else besides. I speak about all of this in Chapter 8.

There is one more observation which I would like to make

* A sphere is the set of points in R^3 determined by the equation $x^2 + y^2 + z^2 = 1$. The 3-sphere is the set of points in R^4 determined by the equation $x^2 + y^2 + z^2 + u^2 = 1$.

about Einstein's 3-sphere. However incredible it might seem, the same idea had already been conceived by another genius, from an entirely different cultural universe: Dante Alighieri, Italy's greatest poet. In the *Paradiso*, the third part of his major poem, the *Commedia*, Dante offers a grandiose vision of the medieval world, calqued on the world of Aristotle, with the spherical Earth at its centre, surrounded by the celestial spheres (figure 3.13).

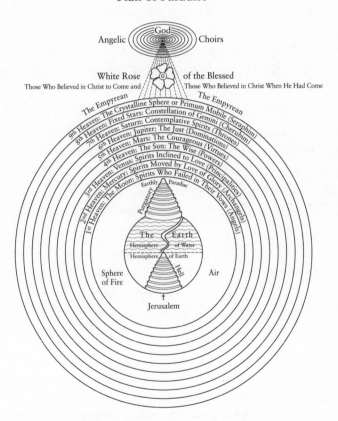

Plan of Paradise

Angelic — God — Choirs

White Rose of the Blessed
Those Who Believed in Christ to Come and / \ Those Who Believed in Christ When He Had Come

The Empyrean   The Empyrean
9th Heaven: The Crystalline Sphere or Primum Mobile (Seraphim)
8th Heaven: Fixed Stars: Constellation of Gemini (Cherubim)
7th Heaven: Saturn: Contemplative Spirits (Thrones)
6th Heaven: Jupiter: The Just (Dominations)
5th Heaven: Mars: The Courageous (Virtues)
4th Heaven: The Sun: The Wise (Powers)
3rd Heaven: Venus: Spirits Inclined to Love (Principalities)
2nd Heaven: Mercury: Spirits Moved by Love of Glory (Archangels)
1st Heaven: The Moon: Spirits Who Failed in Their Vows (Angels)

Earthly Paradise

Purgatory

The Earth
Hemisphere of Water
Hemisphere of Earth

Hell

Sphere of Fire

Air

† Jerusalem

Figure 3.13 Traditional representation of Dante's universe.

81

Accompanied by his shining loved one, Beatrice, Dante ascends these spheres in the course of a fantastic, visionary journey up to the outermost sphere. When he reaches it, he contemplates the universe below him with its rotating heavens and the Earth, very far down, at its centre. But then he looks even higher – and what does he see? He sees a point of light surrounded by immense spheres of angels, that is to say, by another immense ball, which, in his words 'surrounds and is at the same time surrounded by' the sphere of our universe! Here are Dante's verses from Canto XXVII of the *Paradiso. Questa altre parte dell'Universo d'un cerchio lui comprende si come questo li altri*: 'This other part of the universe surrounds the first in a circle like the first surrounds the others.' And in the next canto, still on the last 'circle', *parendo inchiuso da quel ch'elli inchiude*: 'appearing to be to be enclosed by those that it encloses'. The point of light and the sphere of angels are surrounding the universe, and at the same time they are *surrounded by* the universe! It is an exact description of a 3-sphere!

The usual representations of Dante's universe common in Italian schoolbooks (such as figure 3.13) place the angelic spheres separate from the celestial ones. But Dante writes that the two balls 'surround and are surrounded by' each other. Dante has a clear geometrical intuition of a 3-sphere.*

The first to notice that the *Paradiso* describes the universe as a 3-sphere was an American mathematician, Mark Peterson, in 1979. In general, scholars of Dante are not very familiar with 3-spheres. Today, every physicist and mathematician could

---

* It has been objected that Dante speaks of 'circles' and not of 'spheres'. But the objection is invalid. Brunetto Latini writes of 'a circle, like the shell of an egg'. The word 'circle', for Dante, as for his teacher and mentor, designates everything which is circular, including spheres.

easily recognize the 3-sphere in Dante's description of the universe.

How is it possible that Dante had an idea that sounds so modern? I think it was possible, in the first place, due to the profound intelligence of Italy's finest poet. This intelligence is one of the reasons why the *Commedia* is so fascinating. But it is also due to the fact that Dante was writing well before Newton convinced everyone that the infinite space of the cosmos was the flat one of Euclidean geometry. Dante was free of the restraints upon our intuition we have as a result of our Newtonian schooling.

Dante's scientific culture was based principally on the teachings of his mentor and tutor, Brunetto Latini, who has left us a small, enchanting treatise, *Li tresor*, which is a sort of encyclopaedia of medieval knowledge, written in a delightful combination of old French and Italian. In *Li tresor*, Brunetto explains in detail the fact that the Earth is round. But he does so, curiously to the eyes of a modern reader, in terms of 'intrinsic' rather than 'extrinsic' geometry. That is to say, he does not write, 'the Earth is like an orange', as the Earth would look if seen from the outside, but writes instead, 'Two knights who could gallop sufficiently far in opposite directions would meet up on the other side.' And: 'If he were not impeded by the seas, a man who set out to walk for ever would return to the point on the Earth from which he departed.' In other words, he adopts an internal, not an external, point of view: the perspective of someone who walks the Earth, not of someone who looks at it from afar. At first glance it might seem like a pointless, complicated way of explaining that the Earth is a ball. Why doesn't Brunetto simply say that the Earth is like an orange? But, on reflection: if, say, an ant walks on an orange, it will at some point find itself upside down, and must keep itself attached by means of the tiny suction pads on its legs, to avoid falling off. And yet a traveller who walks the Earth never finds himself upside down, and needs no suction pads on his legs. Brunetto's description is not so quaint after all.

Now, think about it. For someone who has learned from his teacher that the form of the surface of our planet is such that by walking always in a straight line we return to the point we started from, it is perhaps not so difficult to take the next obvious step, and imagine that the form of the entire universe is such that, flying always in a straight line, we return to the same point of departure: a 3-sphere is a space in which 'two winged knights that could *fly* in opposite directions would meet up on the other side'. In technical terms, the description of the geometry of the Earth offered by Brunetto Latini in *Li tresor* is given in terms of intrinsic geometry (seen from the inside) rather than extrinsic (seen from the outside), and this is exactly the description that is suitable to generalize the notion of 'sphere' from two dimensions to three. The best way of describing a 3-sphere is not to try to 'see it from the outside', but rather to describe what happens when moving within it.

The method developed by Gauss to describe curved surfaces, and generalized by Riemann to describe the curvature of spaces in three or more dimensions, basically, amounts to Brunetto Latini's way. That is to say, the idea is to describe a curved space not as 'seen from the outside', stating how it curves in an external space, but instead in terms of what may be experienced by somebody *within* that space, who is moving and always remaining within it. For instance, the surface of an ordinary sphere, as Brunetto observes, is a surface where all the 'straight' lines get back to the starting point after traversing the same distance (the length of the equator). A 3-sphere is a three-dimensional space with the same property.

Einstein's spacetime is not curved in the sense that it curves 'in an external space'. It is curved in the sense that its *intrinsic* geometry, that is to say, the web of distances between its points, which can be observed by staying within it, is not the geometry of a flat space. It is a space where Pythagoras's theorem is not valid, just as Pythagoras's theorem is not valid on the surface of the Earth.*

There is a way of understanding the curvature of space from within it, and without looking at it from outside, which is important for what follows. Imagine you are at the North Pole and walk southwards until you reach the equator, carrying with you an arrow pointing ahead. Once you reach the equator, turn to the left without changing the direction of the arrow. The arrow still points south, which is now to your right. Advance a little towards the east along the equator and then turn again towards the north – again without changing the direction of the arrow, which will now be pointing behind you. When you reach the North Pole again, you have executed a closed circuit – a 'loop', as it is termed – and the arrow does not point in the same direction as when you started out (figure 3.14). The angle through which the arrow has turned in the course of the loop measures the curvature.

I will return later to this method of measuring curvature by making a loop in space. These will be the loops that give the name to the theory of loop quantum gravity.

* On the surface of the Earth, for instance, the North Pole and two points on the equator can make a triangle with three sides of equal length and three right angles – something which clearly cannot be done on a plane.

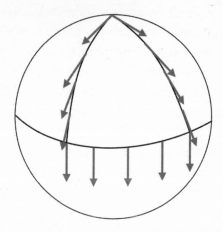

Figure 3.14 An arrow carried parallel to itself along a circuit (a loop) in a curved space arrives back rotated at the point of departure.

Dante leaves Florence in 1301, while the mosaics in the cupola of the Baptistery are being completed. The mosaic, representing Hell (the work of Coppo di Marcovaldo, the teacher of Cimabue), probably terrifying in the eyes of a medieval person, has often been indicated as a source of inspiration to Dante (figure 3.15).

Shortly before starting to write this book, I visited

Figure 3.15 The mosaic depicting Hell, by Coppo di Marcovaldo, in the Baptistery of Florence.

the Baptistery in the company of Emanuela Minnai, the friend who convinced me to write it. Entering the Baptistery and looking up, you see a shining point of light (the light source from the lantern at the summit of the cupola) surrounded by nine orders of angels, with the name of each order written: Angels, Archangels, Principalities, Powers, Virtues, Domains, Thrones, Cherubim and Seraphim. This corresponds exactly to the structure of the second sphere of Paradise. Imagine that you are an ant on the floor of the Baptistery and are able to walk in any direction; regardless of which direction you follow to climb the wall, you would reach the ceiling at the same point of light surrounded by angels: the point of light and its angels both 'surround' and 'are surrounded by' the rest of the decorated interior of the Baptistery (figure 3.16).

Like every citizen of Florence at the end of the thirteenth century, Dante must have been profoundly awe-struck by the Baptistery, the grandiose architectural enterprise his city was

Figure 3.16 The interior of the Baptistery.

completing. I believe that he may have been inspired by the Baptistery, not only by Coppo di Marcovaldo's *Inferno*, but also by its overall architecture, for his vision of the cosmos. The *Paradiso* reproduces its structure remarkably precisely, including the nine circles of angels and the point of light, just translating it from two to three dimensions. After describing the spherical universe of Aristotle, Brunetto had already written that beyond it lies the place of divinity – and medieval iconography had already imagined Paradise as God surrounded by spheres of angels. In the end, Dante does no more than mount the pieces that already existed into a coherent architectural whole which follows the suggestive architecture of the Baptistery and resolves the ancient problem of the borders of the universe. In so doing, Dante anticipates by six centuries Einstein's 3-sphere.

I don't know if the young Einstein had encountered the *Paradiso* during his intellectual wanderings in Italy, and whether or not the vivid imagination of the Italian poet may have had a direct influence on his intuition that the universe might be both finite and without boundary. Whether or not such influence occurred, I believe that this example demonstrates how great science and great poetry are both visionary, and may even arrive at the same intuitions. Our culture is foolish to keep science and poetry separated: they are two tools to open our eyes to the complexity and beauty of the world.

Dante's 3-sphere is only an intuition within a dream. Einstein's 3-sphere has mathematical form and follows from the theory's equations. The effect of each is different. Dante moves us deeply, touching the sources of our emotions. Einstein opens a road towards the unsolved mysteries of our universe. But both count among the most beautiful and significant flights that the mind can achieve.

But let's return to 1917, when Einstein tries to insert the idea of the 3-sphere into his equations. Here he encounters a problem. He is convinced that the universe is fixed and immutable, but his equations tell him that this is not possible. It isn't difficult to understand why. Everything attracts, therefore the only way for a finite universe not to collapse on itself is for it to be expanding: just as the only way to prevent a football from falling to the ground is to kick it upwards. It either goes up, or falls down – it can't stay still, suspended in the air.

But Einstein does not believe what his own equations are telling him. He even makes a silly physics mistake (he does not realize that the solution he considers is unstable) just to avoid accepting what his theory predicts: the universe is either contracting or expanding. He modifies his equations, trying to avoid the implication that it is expanding. It is for this reason that he adds the term Λg_{ab} in the equation written above. But it is a further mistake: the added term is correct, but it does not change the fact that the equation predicts that the universe must be expanding. For all his bravery, Einstein the genius lacks the courage to believe his own equations.

A few years later Einstein is forced to give up: it is his theory that is right, not his reservations about it. Astronomers realize that all galaxies are indeed moving away from us. The universe is expanding, exactly as the equations predicted. Fourteen billion years ago, the universe was concentrated almost to a single, furiously hot point. From there it expanded in a colossal 'cosmic' explosion – and here the term 'cosmic' is not used in any rhetorical sense: it is, literally, a cosmic explosion. This is the 'Big Bang'.

Today we know the expansion is real. The definitive proof of the scenario foreseen by Einstein's equations arrives in

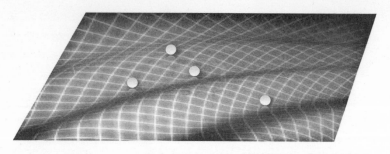

Figure 3.17 Einstein's world: particles and fields
which move on other fields.

1964, when two American radio-astronomers, Arno Penzias
and Robert Wilson, discover by accident a radiation diffused
throughout the universe which turns out to be precisely what
remains of the original immense heat of the early universe.
Once again, the theory turns out to have been correct, up
to its most amazing predictions.

Ever since we discovered that the Earth is round and turns
like a mad spinning-top, we have understood that reality is
not what it seems: every time we glimpse a new aspect of it,
it is a deeply emotional experience. Another veil has fallen.
But the leap made by Einstein is unparalleled: spacetime is a
field; the world is made only of fields and particles; space
and time are not something else, something different
from the rest of nature: they are just a field among the others
(figure 3.17).

In 1953, a primary schoolchild writes to Albert Einstein,
'Our class is studying the universe. I am very interested in
space. I would like to thank you for all that you have done so
that we might understand it.'[5]

I feel the same way.

4. Quanta

The two pillars of twentieth-century physics – general relativity and quantum mechanics – could not be more different from each other. General relativity is a compact jewel: conceived by a single mind, based on combining previous theories, it is a simple and coherent vision of gravity, space and time. Quantum mechanics, or quantum theory, on the other hand, emerges from experiments in the course of a long gestation over a quarter of a century, to which many have contributed; achieves unequalled experimental success and leads to applications which have transformed our everyday lives (the computer on which I write, for instance); but, more than a century after its birth, it remains shrouded in obscurity and incomprehensibility.

This chapter illustrates the strange physics of this theory, relates how the theory came into being and the three aspects of reality it has unveiled: granularity, indeterminism and relationality.

Albert again

It's said that quantum mechanics was born precisely in 1900, virtually ushering in a century of intense thought. In 1900 the German physicist Max Planck tries to compute the amount of electromagnetic waves in equilibrium in a hot box. To obtain a formula reproducing the experimental results, he ends up using a trick which does not appear to make much sense: he assumes that the energy of the electric field is distributed in 'quanta', that is to say, in small packets, little bricks of energy. The size of the packets, he assumes, depends on the frequency (that is, the colour) of the electromagnetic waves. For waves of frequency ν, every quantum, or every packet, has energy

$$E = h\nu$$

This formula is the first of quantum mechanics; h is a novel constant which today we call the Planck constant. It fixes how much energy there is in each packet of energy, for radiation of frequency (colour) ν. The constant h determines the scale of all quantum phenomena.

The idea that energy could be made up of finite packets is at odds with everything that was known at the time: energy was considered something that could vary in a continuous manner, and there was no reason to treat it as if it were made up of grains. For example, the energy of a pendulum measures the amplitude of the swing. There seems to be no reason for a pendulum to oscillate only with certain determined amplitudes and not others. For Max Planck, taking energy in finite-size packets was only a strange trick which happened to work for the calculation – that is, to reproduce laboratory measurements – but for utterly unclear reasons.

Five years later it is Albert Einstein – him again – who comes to understand that Planck's packets of energy are in fact real. This is the subject of the third of the three articles sent to the *Annalen der Physik* in 1905. And this is the true date of birth of quantum theory.

In the article, Einstein argues that light truly is made up of small grains, particles of light. He considers a phenomenon that had been recently observed: the photoelectric effect. There are substances that generate a weak electric current when struck by light. That is to say, they emit electrons when light shines on them. Today we use them, for example, in the photoelectric cells which open doors when we approach them by detecting if light arrives, or not, in a sensor. That this happens is not strange, because light carries energy (it warms us, for example), and its energy makes the electrons 'jump out' of their atoms; it gives them a push.

But something is strange: it seems reasonable to expect that if the energy of light is scarce – namely, if the light is dim – the phenomenon would not take place; and that it would take place when the energy is sufficient – namely, when the light is bright. But it isn't like this: what is observed is that the phenomenon happens only if the *frequency* of light is high and does not happen if the *frequency* is low. That is to say, it happens or doesn't happen depending on the colour of light (the frequency) rather than its *intensity* (energy). There is no way of making sense of this with standard physics.

Einstein uses Planck's idea of the packets of energy, with a size that depends upon frequency, and realizes that if these packets are real, the phenomenon can be explained. It isn't difficult to understand why. Imagine that the light arrives in the form of grains of energy. An electron will be swept out of its atom if the individual grain hitting it has a great deal of energy. What matters is the energy of *each* grain, not the *number* of grains. If, as in Planck's hypothesis, the energy of each grain is determined by frequency, the phenomenon will occur only if frequency is sufficiently high, that is to say, if the *individual* grains of energy are sufficiently large, independently from the total amount of energy that's around.

It is like when it hails: what determines whether your car

will be dented is not the total quantity of hail that falls but the size of the individual hailstones. There could be an enormous amount of hail, but it will do no damage if all the stones are small. In the same way, even if light is intense – which amounts to saying that there are a great deal of light packets – the electrons would not be extracted from their atoms if the individual grains of light are too small, that is, if the frequency of light is too low. This explains why it is the colour and not the intensity which determines whether the photoelectric effect occurs or not. For this simple reasoning Einstein was awarded the Nobel Prize. It is easy to understand things once someone has thought them through. The difficulty lies in thinking them through in the first place.

Today we call these packets of energy 'photons', from the Greek word for light: φώς. Photons are the grains of light, its 'quanta'. In the article Einstein writes:

> It seems to me that the observations associated with black-body radiation, fluorescence, the production of cathode rays by ultraviolet light, and other related phenomena connected with the emission or transformation of light are more readily understood if one assumes that the energy of light is discontinuously distributed in space. In accordance with the assumption to be considered here, the energy of a light ray spreading out from a point source is not continuously distributed over an increasing space but consists of a finite number of 'energy quanta' which are localized at points in space, which move without dividing, and which can only be produced and absorbed as complete units.[1]

These simple and clear lines are the real birth certificate of quantum theory. Note the wonderful initial 'It seems to me . . .', which recalls the hesitations of Faraday, or those of Newton; or the uncertainty of Darwin in the first pages of *On the Origin of Species*. True genius is aware of the momentousness of the steps it is taking, and is always hesitant . . .

There is a clear relation between Einstein's work on Brownian motion (discussed in Chapter 1) and his work on the quanta of light, both completed in 1905. In the first, Einstein had managed to find a demonstration of the atomic hypothesis, that is to say, of the granular structure of matter. In the second he extends this same hypothesis to light: light must have a granular structure as well.

At first, Einstein's idea that light could be made up of photons is regarded by his colleagues as no more than youthful waywardness. Everyone commends him for his theory of relativity, but everybody judges the notion of photons to be outlandish. Scientists had only recently been persuaded that light was a wave in the electromagnetic field: how could it be made up of grains? In a letter addressed to the German Ministry, recommending that Einstein should have a professorship inaugurated for him in Berlin, the most distinguished physicists of the day write that the young man is so brilliant that he 'may be excused' certain excesses, such as the idea of photons. Not many years later, the very same colleagues award him the Nobel Prize, precisely for having understood that photons exist. Light falls on a surface like a gentle hail shower.

To comprehend how light may be simultaneously an electromagnetic wave and a swarm of photons will require the entire construction of quantum mechanics. But the first building block of this theory has been established: there exists a fundamental granularity in *all* things, including light.

Niels, Werner and Paul

If Planck is the biological father of the theory, Einstein is the parent who gave birth to and nurtured it. But as is often the case with children, the theory then went its own way, barely recognized by Einstein as his own.

During the first two decades of the twentieth century, it is the Dane Niels Bohr who is responsible for guiding its development. Bohr studies the structure of atoms, which was beginning to be explored at the turn of the century. Experiments had shown that an atom is like a small solar system: the mass is concentrated in a heavy central nucleus, around which light electrons revolve, more or less like the planets around the Sun. This picture, however, did not account for a simple fact: matter is coloured.

Salt is white, pepper is black, chilli is red. Why? Studying the light emitted by atoms, it is apparent that substances have specific colours. Since colour is the frequency of light, light is emitted by substances at certain fixed frequencies. The set of the frequencies that characterizes a given substance is known as the 'spectrum' of this substance. A spectrum is a collection of fine lines of different hues, in which the light emitted by a given substance is decomposed (for instance, by a prism). The spectra of a few elements are shown in figure 4.2.

Figure 4.1 Niels Bohr.

Spectra of numerous substances had been studied and catalogued in many laboratories

96

at the turn of the century, and nobody knew how to explain why each substance had this or that spectrum. What determines the colour of those lines?

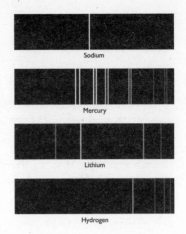

Colour is the speed at which Faraday's lines vibrate, and this is determined by the vibrations of the electric charges which emit light. These charges are the electrons that move inside the atoms. Therefore, studying spectra, we can understand how electrons move around nuclei. The other way around, we could predict the spectrum of each atom by computing the frequencies of the electrons circling their nucleus. Easy to say, but in practice nobody was able to do so. In fact, the whole thing seemed impossible, because in Newton's mechanics an electron can revolve around its nucleus at *any* speed, and hence emit light at *any* frequency. But then why does the light emitted by an atom not contain *all* colours, rather than just a few particular ones? Why are atomic spectra not a continuum of colours, instead of just a few separate lines? Why, in technical parlance, are they 'discrete' instead of continuous? For decades, physicists seemed incapable of finding an answer.

Bohr finds a tentative solution, by way of a strange hypothesis. He realizes that everything could be explained if the energy of electrons in atoms could only assume certain 'quantized' values – certain specific values, just as was hypothesized by Planck and by Einstein for the energy of the quanta

Figure 4.2 The spectra of some elements: sodium, mercury, lithium and hydrogen.

of light. Once again, the key is a *granularity*, but not now for the energy of light but rather for the energy of the electrons in the atom. It begins to become clear that granularity is something widespread in nature.

Bohr makes the hypothesis that electrons can exist only at certain 'special' distances from the nucleus, that is, only on certain particular orbits, the scale of which is determined by Planck's constant h. And that electrons can 'leap' between one orbit with the permitted energy to another. These are the famous 'quantum leaps'. The frequency at which the electron moves on these orbits determines the frequency of the emitted light and, since only certain orbits are allowed, it follows that only certain frequencies are emitted.

These hypotheses define Bohr's 'atomic model', whose centenary was commemorated in 2013. With these assumptions (outlandish, but simple) Bohr manages to compute the spectra of all atoms, and even to predict accurately spectra not yet observed. The experimental success of this simple model is astonishing.

Clearly, there must be some truth in these assumptions, even if they run contrary to all contemporary notions of matter and dynamics. But why are there always only just certain orbits? And what does it mean to say that electrons 'leap'?

In Bohr's institute in Copenhagen, the most brilliant young minds of the century gather to try to give order to this jumble of incomprehensible behaviours in the atomic world, and to construct a coherent theory. The research is arduous and protracted, until a young German finds the key to unlock the door of the mystery of the quantum world.

Werner Heisenberg is twenty-five years old when he writes the equations of quantum mechanics, the same age as

Einstein was when he wrote his three major articles. He does so on the basis of dizzying ideas.

The intuition comes to him one night in the park behind the Copenhagen Institute of Physics. The young Werner walks about pensively in the park. It is really dark there; we are in 1925. There is only an occasional streetlamp, casting dim islands of light here and there. The pools of light are separated by large expanses of darkness. Suddenly, Heisenberg

Figure 4.3 Werner Heisenberg.

sees a figure pass by. Actually, he does not see him pass: he sees him appear beneath a lamp, then disappear into the dark before reappearing beneath another lamp, and then vanishing back into the dark again. And so on, from pool of light to pool of light, until he eventually disappears altogether into the night. Heisenberg thinks that, 'evidently', the man does not actually vanish and reappear: in his mind, he can easily reconstruct the man's trajectory between one streetlamp and another. After all, a man is a substantial object, big and heavy – and big, heavy objects do not simply appear and vanish . . .

Ah! *These* objects, which are substantial, large and heavy, don't vanish and reappear . . . but what do we know about electrons? A light flashes on in his mind. Why should small objects such as electrons do the same? What if, effectively, electrons *could* vanish and reappear? What if these were the mysterious quantum leaps which appeared to underlie the structure of the atomic spectra? What if, between one interaction with something, and another with something else, the electron could literally be *nowhere*.

What if the electron could be something that manifests itself only when it interacts, when it collides with something else; and that between one interaction and another it had no precise position? What if always having a precise position is something which is acquired only if one is substantial enough – large and heavy like the man that passed by a little while ago, like a ghost in the dark, and then disappeared into the night . . . ?

Only someone in his twenties can take such delirious propositions seriously. You have to be a twenty-something to believe that they can be turned into a theory of the world. And perhaps you have to be this young to understand better than anyone else, for the first time, the deep structure of nature. Just as Einstein was in his twenties when he realized that time does not pass in the same way for everyone, so, too, was Heisenberg on that Copenhagen night. Perhaps, it is no longer a good idea to trust your intuitions after the age of thirty . . .

Heisenberg returns home gripped by feverish emotion, and plunges into calculations. He emerges, some time later, with a disconcerting theory: a fundamental description of the movement of particles, in which they are described not by their position at every moment but only by their position at particular instants: the instants in which they interact with something else.

This is the second cornerstone of quantum mechanics, its hardest key: the *relational* aspect of things. Electrons don't always exist. They exist when they interact. They materialize in a place when they collide with something else. The quantum leaps from one orbit to another constitute their way of being real: an electron is a combination of leaps from one interaction to another. When nothing disturbs it, an electron does not exist in any place. Instead of writing the position and velocity of the electron, Heisenberg writes tables of numbers (matrices). He multiplies and divides tables of numbers representing possible interactions of the electron. And, as if from the magical abacus of a magus, the results correspond exactly with what was observed. These are the first fundamental equations of quantum mechanics. From here on, these equations will do nothing but work, work, work. Up until now, incredible as it may seem, they have never failed.

In the end, it is another twenty-five-year-old who picks up the work initiated by Heisenberg, takes the new theory in his hands and constructs its entire formal and mathematical scaffolding: the Englishman Paul Adrien Maurice Dirac, considered by many to be the greatest physicist of the twentieth century after Einstein.

Despite his scientific stature, Dirac is much less well-known than Einstein. This is due, in part, to the rarefied abstraction of his science, and partly due to his disconcerting character. Silent in company, extremely reserved, incapable of expressing emotions, frequently unable

Figure 4.4 Paul Dirac.

to recognize the faces of acquaintances – incapable even of conducting an ordinary conversation, or of, apparently, understanding simple questions – he seemed virtually autistic, and perhaps fell within the spectrum of this condition.

During one of his lectures, a colleague said to him, 'I don't understand that formula.' After a short, silent pause, Dirac continued on regardless. The moderator interrupted him, asking if he would like to reply to the question. Dirac, sincerely astonished, replied, 'Question? What question? My colleague has made an assertion.' And so, in a very pedantic sense, he had. It wasn't arrogance: the man who could discover secrets of nature which had eluded everyone else could not understand the implicit meaning of language, could not grasp its non-literal usage, and took every phrase at face value.[2] And yet, in his hands, quantum mechanics is transformed from a jumble of intuitions, half-baked calculations, misty metaphysical discussions and equations that work well, but inexplicably, into a perfect architecture: airy, simple and extremely beautiful. Beautiful, but stratospherically abstract.

The venerable Bohr said of him, 'Of all physicists, Dirac has the purest soul.' And don't his eyes, in figure 4.4 show so? His physics has the pristine clarity of a song. For him, the world is not made of things, it's constituted of an abstract mathematical structure which shows us how things appear and how they behave when manifesting themselves. It's a magical encounter between logic and intuition. Deeply impressed, Einstein remarked, 'Dirac poses problems for me. To maintain an equilibrium along this vertiginous course, between genius and madness, is a daunting enterprise.'

Dirac's quantum mechanics is the mathematical theory used today by any engineer, chemist or molecular biologist. In it, every object is defined by an abstract space* and has no property in itself, apart from those that are unchanging, such as mass. Its position and velocity, its angular momentum and its electrical potential, and so on, acquire

* A Hilbert space.

reality only when it collides – 'interacts'– with another object. It is not just its position which is undefined, as Heisenberg had recognized: *no* variable of the object is defined between one interaction and the next. The *relational* aspect of the theory becomes universal.

When it suddenly appears, in the course of an interaction with another object, a physical variable (velocity, energy, momentum, angular momentum) does not assume just any value. Dirac provides the general recipe to compute the set of values that a physical variable can take.* These values are analogous to the spectra of the light emitted by atoms. Today we call the set of the particular values which a variable may assume the 'spectrum' of that variable, by analogy with the spectra into which the light of elements decomposes – the first manifestation of this phenomenon. For example, the radius of the orbitals of an electron around a nucleus can acquire only specific values, those that Bohr had hypothesized, which form the 'spectrum of the radius'.

The theory also gives information on *which* value of the spectrum will manifest itself in the next interaction, but only in the form of probabilities. We do not know with certainty where the electron will appear, but we can compute the *probability* that it will appear here or there. This is a radical change from Newton's theory, where it is possible, in principle, to predict the future with certainty. Quantum mechanics brings probability to the heart of the evolution of things. This *indeterminacy* is the third cornerstone of quantum mechanics: the discovery that chance operates at the atomic level. While Newton's physics allows for the prediction of the future with

* These are the eigenvalues of the operator associated with the physical variable in question. The key equation is the eigenvalue equation.

exactitude, if we have sufficient information about the initial data and if we can make the calculations, quantum mechanics allows us to calculate only the *probability* of an event. This absence of determinism at a small scale is intrinsic to nature. An electron is not obliged by nature to move towards the right or the left; it does so by chance. The apparent determinism of the macroscopic world is due only to the fact that the microscopic randomness cancels out on average, leaving only fluctuations too minute for us to perceive in everyday life.

Dirac's quantum mechanics thus allows us to do two things. The first is to calculate *which* values a physical variable may assume. This is called 'calculation of the spectrum of a variable'; it captures the *granular* nature of things. When an object (atom, electromagnetic field, molecule, pendulum, stone, star, and so on) interacts with something else, the values computed are those which its variables can assume in the interaction (*relationism*). The second thing that Dirac's quantum mechanics allows us to do is to compute the *probability* that this or that value of a variable appears at the next interaction. This is called 'calculation of an *amplitude of transition*'. Probability expresses the third feature of the theory: *indeterminacy* – the fact that it does not give unique predictions, only probabilistic ones.

This is Dirac's quantum mechanics: a recipe for calculating the spectra of the variables and a recipe for calculating the probability that one or another value in the spectrum will appear during an interaction. That's it. What happens between one interaction and the next is not mentioned in the theory. It does not exist.

The probability of finding an electron or any other particle at one point or another can be imagined as a diffuse cloud, denser where the probability of seeing the particle is stronger. Sometimes it is useful to visualize this cloud as if it were a real thing. For instance, the cloud that represents an electron around its nucleus indicates where it is more likely that the electron appears if we look at it. Perhaps you encountered them at school: these are the atomic 'orbitals'.*

The efficacy of the theory soon proves extraordinary. If today we build computers, have advanced molecular chemistry and biology, lasers and semiconductors, it is thanks to quantum mechanics. For a certain number of decades it was as if it were Christmas every day for physicists: for every new

---

* This cloud is described by a mathematical object called wave function. The Austrian physicist Erwin Schrödinger has written an equation describing its evolution in time. Quantum mechanics is often mistakenly identified with this equation. Schrödinger had hopes that the 'wave' could be used to explain the oddities of quantum theory: from those of the sea to electromagnetic ones, waves are something we understand well. Even today, some physicists try to understand quantum mechanics by thinking that reality is the Schrödinger wave. But Heisenberg and Dirac understood at once that this would not do. To view Schrödinger's wave as something real is to give it too much weight – it doesn't help us to understand the theory; on the contrary, it leads to greater confusion. Except for special cases, the Schrödinger wave is not in physical space, and this divests it of all its intuitive character. But the main reason why Schrödinger's wave is a bad image of reality is the fact that, when a particle collides with something else, it is always at a point: it is never spread out in space like a wave. If we conceive an electron as a wave, we get in trouble explaining how this wave instantly concentrates to a point at each collision. Schrödinger's wave is not a useful representation of reality: it is an aid to calculation which permits us to predict with some degree of precision where the electron will reappear. The reality of the electron is not a wave: it is how it manifests itself in interactions, like the man who appeared in the pools of lamplight while the young Heisenberg wandered pensively in the Copenhagen night.

problem, there was an answer which followed from the equations of quantum mechanics, and it was always the correct answer. One example of this will suffice.

The matter surrounding us is made up of a thousand different substances. During the nineteenth and twentieth centuries chemists understood that all these different substances are just combinations of a relatively small number (less than a hundred) of simple elements: hydrogen, helium, oxygen, and so on – to uranium. Mendeleev put these elements in order (according to weight) in the famous periodic table which is pinned to the walls of so many classrooms and which summarizes the properties of the elements of which the world is made – not only on Earth but all over the universe in all galaxies. Why these specific elements? What explains the periodic structure of the table? Why does each element have certain properties and not others? Why, for instance, do some elements combine easily, whereas others do not? What is the secret of the curious structure of Mendeleev's table?

Well, take the equation of quantum mechanics that determines the form of the orbitals of an electron. This equation has a certain number of solutions, and these solutions correspond exactly to hydrogen, helium, oxygen . . . and the other elements! Mendeleev's periodic table is structured exactly like the set of these solutions. The properties of the elements, with everything else, follows from the solution of this equation. Quantum mechanics deciphers perfectly the secret of the structure of the periodic table of elements.

Figure 4.5 Light is a wave on a field, but it has also a granular structure.

Pythagoras and Plato's ancient dream is realized: to describe all of the world's substances with a single formula. The infinite complexity of chemistry, captured by the solutions of a single equation! And this is just one of the applications of quantum mechanics.

Fields and particles are the same thing

Shortly after completing the general formulation of quantum mechanics, Dirac realizes that the theory can be directly applied to fields such as electromagnetic ones, and can be made consistent with special relativity. (Making it consistent with *general* relativity will prove much harder, and is the main subject of this book). In doing this, Dirac discovers an ulterior, profound simplification of our description of nature: the convergence between the notion of particles used by Newton and the notion of fields introduced by Faraday.

The cloud of probability which accompanies electrons between one interaction and another does resemble a field. Faraday and Maxwell's fields, in turn, are made up of grains: photons. Not only are the particles in a certain sense diffused in space like fields but the fields interact like particles. The notions of fields and particles, separated by Faraday and Maxwell, end up merging in quantum mechanics.

The way this happens in the theory is elegant: the equation of Dirac determines the values a variable can take. Applied to the *energy* of Faraday's lines, they tell us that this energy can only take on certain values, and not others. Since the energy of the electromagnetic field can take on only certain values, the field behaves like a set of packets of energy. These are precisely the quanta of energy introduced by Planck and

Einstein thirty years earlier. The circle closes, and the story is complete. The equations of the theory, written by Dirac, account for the granular nature of light, which Planck and Einstein had intuited.

The electromagnetic waves are vibrations of Faraday's lines, but also, at a small scale, swarms of photons. When they interact with something else, as in the photoelectric effect, they manifest themselves as particles: to our eye, light rains in separate droplets, in single photons. Photons are the quanta of the electromagnetic field.

But the electrons and all the other particles of which the world is made are equally quanta of a field – a 'quantum field' similar to Faraday and Maxwell's, subject to granularity and to quantum probability. Dirac writes the equations for the field of the electrons and of the other elementary particles.* The sharp distinction between fields and particles introduced by Faraday vanishes.

The general form of quantum theory compatible with special relativity is thus called quantum field theory, and it forms the basis of today's particle physics. Particles are quanta of a field, just as photons are quanta of light. All fields display a granular structure in their interactions.

During the course of the twentieth century the list of fundamental fields was repeatedly updated, and today we have a theory called the *standard model* of elementary particles which describes almost all we see, with the exception of gravity,† in the context of quantum field theory. The development of

---

* Dirac's equation.
† There is a phenomenon which seems not to be reducible to the standard model: 'dark matter'. Astrophysicists and cosmologists observe in the universe effects of matter which seems not to be the type of matter

this model occupied physicists for a good part of the last century, and represents in itself a wonderful adventure of discovery. I don't present this side of the story here: it is quantum gravity that I would like to get on to. The standard model is completed by the 1970s. There are approximately fifteen fields, whose quanta are the elementary particles (electrons, quarks, muons, neutrinos, Higgs, and little else), plus a few fields similar to the electromagnetic one, which describe electromagnetic force and the other forces operating at a nuclear scale, whose quanta are similar to the photons.

The standard model was not taken very seriously at first, due to its somewhat cobbled-together aspect, so different from the airy simplicity of general relativity and Maxwell's or Dirac's equations. Against expectations, however, all of its predictions have been confirmed. For more than thirty years, every single experiment of particle physics has done nothing but repeatedly reconfirm the standard model. A recent confirmation was the discovery of the Higgs particle, which caused a sensation in 2013. Introduced to render the theory coherent, the Higgs field seemed a bit artificial – until the Higgs particle, the quantum of this field, was actually observed and found to have precisely the properties predicted by the standard model.* (The fact that it has been

---

described by the standard model. Out there, there are still many things that we don't know.

* I find the claim that the Higgs boson 'explains mass' exaggerated. The Higgs boson does not 'explain' anything about the origin of mass. What would 'explain' the mass of the Higgs? The point is technical: the standard model relies on certain symmetries, and these symmetries seemed to permit only particles devoid of mass. But Higgs and others realized that it is possible to have both symmetries and mass, as long as the latter enters indirectly via the interaction with the field known today as the Higgs field.

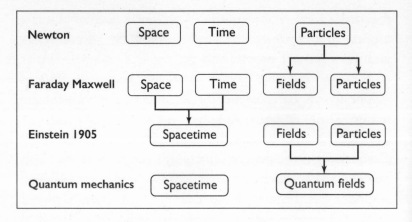

Figure 4.6 What is the world made of?

called 'the God particle' is so stupid as to be unworthy of comment.) In short, despite its unjustly modest name, the standard model has been a triumph.

Quantum mechanics, with its fields/particles, offers today a spectacularly effective description of nature. The world is not made up of fields and particles but of a single type of entity: the quantum field. There are no longer particles which move in space with the passage of time, but quantum fields whose elementary events happen in spacetime. The world is strange, but simple (figure 4.6).

Quanta 1: Information is finite

The time has come to attempt some conclusions about what it is, precisely, that quantum mechanics tells us about the world. It isn't an easy task, because quantum mechanics is not conceptually clear and its true meaning remains controversial; but it's a necessary exercise, to gain clarity and go forward. I think that quantum mechanics has revealed three aspects of the nature of things: granularity, indeterminacy and the relational structure of the world. Let's look at each of these more closely.

The first is the existence of a fundamental *granularity* in nature. The granularity of matter and light is at the heart of quantum theory. It isn't the same granularity intuited by Democritus, however. For Democritus, atoms were like little pebbles, whereas in quantum mechanics particles vanish and reappear. But the root of the idea of the substantive granularity of the world is still to be found in ancient atomism, and quantum mechanics – strengthened by centuries of experiments, by powerful mathematics, and by its extraordinary capacity for making correct predictions – is a genuine recognition of the profound insights on the nature of things reached by the great philosopher of Abdera.

Say we make measurements on a physical system and find that the system is in a particular state. For instance, we measure the amplitude of the oscillations of a pendulum and find that it has a certain value – say, somewhere between five centimetres and six centimetres (no measurement is exact in physics). Before quantum mechanics we would have said that, since there are an infinite number of possible values between five and six centimetres (for instance 5.1 or 5.101 or 5.101001 . . .), then there are *infinite* possible states of motion in which the pendulum could find itself: the amount of our ignorance about the pendulum state is still infinite.

Instead, quantum mechanics tells us that between five and six centimetres there is a *finite* number of possible values of the amplitude, hence our missing information about the pendulum is *finite*.

This goes for everything in general.* Therefore, the first meaning of quantum mechanics is the existence of a limit to the *information* that can exist within a system: a limit to the number of distinguishable states in which a system can be. This limitation upon infinity – this granularity of nature glimpsed by Democritus – is the first central aspect of the theory. Planck's constant h measures the elementary scale of this granularity.

Quanta 2: Indeterminacy

The world is a sequence of granular quantum events. These are discrete, granular and individual; they are individual interactions of one physical system with another. An electron, a quantum of a field or a photon does not follow a trajectory in space but appears in a given place and at a given time when colliding with something else. When and where will it appear? There is no way of knowing with certainty. Quantum mechanics introduces an elementary *indeterminacy* to the heart of the world. The future is genuinely unpredictable. This is the second fundamental lesson learned with quantum mechanics.

Due to this indeterminacy, in the world described by quantum mechanics, things are constantly subject to random change. All the variables 'fluctuate' continually, as if, at the

---

* A finite region of the phase space – the space of the possible states of a system – contains an *infinite* number of distinguishable classic states, but *always* only a *finite* number of orthogonal quantum states. This number is given by the volume of the region, divided by the Planck constant, raised to the number of degrees of freedom. This result is general.

smallest scale, everything is constantly vibrating. We do not see these omnipresent fluctuations only because of their small scale; they cannot be observed at a large scale, as when we observe macroscopic bodies. If we look at a stone, it stays still. But if we could see its atoms, we would observe them constantly spread here and there, and in ceaseless vibration. Quantum mechanics reveals to us that, the more we look at the detail of the world, the less constant it is. The world is not made up of tiny pebbles. It is a world of vibrations, a continuous fluctuation, a microscopic swarming of fleeting micro-events.

The atomism of antiquity had anticipated also this aspect of modern physics: the appearance of laws of probability at a deep level. Democritus assumed (just like Newton) that the movement of atoms was rigorously determined by their collisions. But his successor, Epicurus, corrects the determinism of the master and introduces into atomism the notion of indeterminacy – in the same way in which Heisenberg introduces indeterminacy into Newton's determinism. For Epicurus, atoms can on occasion deviate by chance from their course. Lucretius says this in beautiful words: this deviation occurs '*incerto tempore . . . incertisque loci*':[3] at an uncertain place, at an uncertain time. The same randomness, the same appearance of probability at an elementary level, is the second key discovery about the world that quantum mechanics expresses.

So, how do we compute the probability that an electron in a certain initial position A will reappear, after a given time, in one or another final position B?

In the 1950s, Richard Feynman, who I've already mentioned, found a suggestive method of making this calculation: consider *all* possible trajectories from A to B, that is to say,

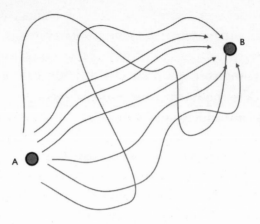

Figure 4.7 In order to move from A to B an electron
behaves as if passing through *all* possible trajectories.

all possible trajectories the electron can follow (straight,
curved, zigzagging . . .). Each trajectory determines a number. The probability is obtained from the sum of all these
numbers. The details of this calculation are not important:
what matters is the fact that all trajectories from A to B
contribute: it is as if the electron, in order to go from A to
B, passed 'through all possible trajectories', or, in other
words, unfurled into a cloud in order then to converge mys-
teriously on point B, where it collides again with something
else (figure 4.7).

This technique for computing the probability of a quantum
event is called Feynman's sum over paths,* and we shall see
that it plays a role in quantum gravity.

---

* Or Feynman's integral. The probability of going from A to B is the
square module of the integral over all the paths of the exponential of

Quanta 3: Reality is relational

The third discovery about the world articulated by quantum mechanics is the most profound and difficult – and one which was not anticipated by the atomism of antiquity.

The theory does not describe things as they *are*: it describes how things *occur* and how they *interact with each other*. It doesn't describe where there is a particle but how the particle *shows itself to others*. The world of existent things is reduced to a realm of possible interactions. Reality is reduced to interaction. Reality is reduced to relation.[4]

In a certain sense, this is just an extension of relativity, albeit a radical one. Aristotle was first to emphasize that we only perceive *relative* speed. On a ship, for example, we talk of our speed relative to the ship; on land, relative to the Earth. Galileo understood that this is the reason why the Earth can move with respect to the Sun without us feeling the movement. Speed is not a property of an object on its own: it is the property of the motion of an object *with respect to another object*. Einstein extended the notion of relativity to time: we can say that two events are simultaneous only relatively to a given motion (see footnote 2 in Chapter 3 above). Quantum mechanics extends this relativity in a radical way: *all* variable aspects of an object exist only in relation to other objects. It is only in interactions that nature draws the world.

In the world described by quantum mechanics there is no reality except in the *relations* between physical systems. It isn't

---

the classical action of the trajectory, multiplied by the imaginary unit and divided by Planck's constant.

things that enter into relations but, rather, relations that ground the notion of 'thing'. The world of quantum mechanics is not a world of objects: it is a world of events. Things are built by the happening of elementary events: as the philosopher Nelson Goodman wrote in the 1950s, in a beautiful phrase, 'An object is a monotonous process.' A stone is a vibration of quanta that maintains its structure for a while, just as a marine wave maintains its identity for a while before melting again into the sea.

What is a wave, which moves on water without carrying with it any drop of water? A wave is not an object, in the sense that it is not made of matter that travels with it. The atoms of our body, as well, flow in and away from us. We, like waves and like all objects, are a flux of events; we are processes, for a brief time monotonous . . .

Quantum mechanics does not describe objects: it describes processes and events which are junction points between processes.

To summarize, quantum mechanics is the discovery of three features of the world:

- *Granularity* (figure 4.8). The information in the state of a system is finite, and limited by Plank's constant.
- *Indeterminacy*. The future is not determined unequivocally by the past. Even the more rigid regularities we see are, ultimately, statistical.
- *Relationality*. The events of nature are always interactions. All events of a system occur in relation to another system.

Quantum mechanics teaches us not to think about the world in terms of 'things' which are in this or that state but in terms of 'processes' instead. A process is the passage from

116

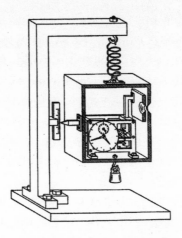

Figure 4.8 The 'light box' in Einstein's mental
experiment, as drawn by Bohr.

one interaction to another. The properties of 'things' manifest themselves in a *granular* manner only in the moment of
interaction, that is to say, at the edges of the processes, and
are such only in *relation* to other things. They cannot be predicted in an unequivocal way but only in a *probabilistic* one.

This is the vertiginous dive taken by Bohr, Heisenberg and
Dirac – into the depth of the nature of things.

But do we really understand?

Certainly, quantum mechanics is a triumph of efficacy. And
yet . . . are you sure, dear reader, that you have fully understood what quantum mechanics reveals to us? An electron is
nowhere when it is not interacting . . . mmm . . . things only
exist by jumping from one interaction to another . . . well . . .
Does it all seem a little absurd?

It seemed absurd to Einstein.

On the one hand, Einstein proposed Werner Heisenberg and Paul Dirac for the Nobel Prize, recognizing that they had understood something fundamental about the world. On the other, he took every opportunity to grumble that, however, none of this made much sense

The young lions of the Copenhagen group were dismayed: how could this come from *Einstein* himself? Their spiritual father, the man who had the courage to think the unthinkable, now pulled back and feared this new leap into the unknown – the very leap which he had himself triggered. How could it be that the same Einstein, who had taught us that time is not universal and that space bends, was now saying that the world could not be *this* strange?

Niels Bohr patiently explained the new ideas to Einstein. Einstein objected. Bohr, in the end, always managed to find answers to the objections. The dialogue continued for years, by way of lectures, letters, articles … Einstein devised mental experiments to show that the new ideas were contradictory: 'Imagine a box filled with light, from which is let escape for a brief instant a single photon . . .': thus one of the most famous examples of these begins (figure 4.8).*

During the course of the exchange, both great men had to

---

* A mechanism in the box opens the small window on the right for an instant, allowing a photon to escape at some precise time. By weighing the box, it is possible to deduce the energy of the released photon. Einstein hoped that this would create difficulties for quantum mechanics, which predicts that time and energy cannot be both precisely determined. Bohr replied, mistakenly, that the way out of the difficulty required Einstein's general relativity, and Einstein, mistakenly, accepted Bohr's reply. The correct response to Einstein, which Bohr was unable to find but that is clear today, is that the position of the escaping photon

give way, to alter their ideas. Einstein was obliged to recognize that there was actually no contradiction within the new ideas. But Bohr had to recognize that things were not as simple and as clear as he thought. Einstein did not want to relent on what for him was the key point: the notion that there is an objective reality, independent of whatever interacted with what. He refused to accept the relational aspect of the theory, the fact that things manifest themselves only through interactions. Bohr did not want to concede on the validity of the profoundly new way in which the real was conceptualized by the theory. Ultimately, Einstein accepts that the theory represents a gigantic leap forward in our understanding of the world, and that it is coherent. But he remains convinced that things could not be as strange as this theory proposed – and that, 'behind' it, there must be a further, more reasonable explanation.

A century has passed, and we are at the same point. Richard Feynman, who more than anyone has known how to juggle with the theory, has written, 'I think I can state that nobody really understands quantum mechanics.'

The equations of the theory and their consequences are used daily in a wide variety of fields: by physicists, engineers, chemists and biologists. But they remain mysterious: they do not describe physical systems but only how physical systems interact with and affect one another. What does this mean?

Physicists and philosophers continue to ask themselves what the real meaning of the theory might be and, in recent years, articles and conferences on the issue have proliferated.

_____

and the weight of the box remain tied to each other ('correlated'), even if the photon is already far away.

What is quantum theory, a century after its birth? An extraordinary dive deep into the nature of reality? A blunder that works, by chance? Part of an incomplete puzzle? Or a clue to something profound regarding the structure of the world, which we have yet to fully decipher?

The interpretation of quantum mechanics which I have presented here is the one which seems least unreasonable to me. It is called the 'relational interpretation', and it has been discussed by serious philosophers such as Bas van Fraassen, Michel Bitbol and Mauro Dorato.[5] But there is no consensus on how to think about quantum mechanics: there are other ways of thinking about it, discussed by other physicists and other philosophers. We are on the brink of that which we don't know, and opinions diverge.

Quantum mechanics is only a physics theory: perhaps tomorrow it will be corrected by an understanding of the world which is different and even more profound. Some scientists today try to iron it out a bit, to render it more in keeping with our intuition. In my opinion, its dramatic empirical success should compel us to take it seriously, and to ask ourselves not what there is to change in the theory – but rather what is limited about our intuition that makes it seem so strange to us.

I think that the obscurity of the theory is not the fault of quantum mechanics but, rather, is due to the limited capacity of our imagination. When we try to 'see' the quantum world, we are rather like moles used to living underground to whom someone is trying to describe the Himalayas. Or like the men imprisoned at the back of Plato's cave.

When Einstein died, his greatest rival, Bohr, found for him words of moving admiration. When, a few years later,

Bohr in turn died, someone took a photograph of the black-board in his study. There's a drawing on it. It represents the 'box of light' of Einstein's thought experiment. To the very last, the desire to debate, to understand more. To the very last, doubt.

This permanent doubt, the deep source of science.

PART THREE

Quantum Space and Relational Time

If you have followed me this far, you now have all the elements with which to understand the current image of the world suggested by fundamental physics – its power, its weaknesses, its limits.

There is a curved spacetime born 14 billion years ago – nobody knows how – and still expanding. This space is a real object, a physical field with its dynamics described by Einstein's equations. Space bends and curves under the weight of matter and plunges into black holes when matter is too concentrated.

Matter is distributed in 100 billion galaxies, each containing 100 billion stars, and is made up of quantum fields which manifest themselves in the form of particles, such as electrons and photons, or as waves, such as the electromagnetic ones that bring us television images and the light of the Sun and the stars.

These quantum fields make up atoms, light and the full contents of the universe. They are strange objects: their quanta are particles that appear when they interact with something else; left alone, they unfurl into a 'cloud of probability'. The world is a swarming of elementary events, immersed in the sea of a vast dynamical space which sways like the water of an ocean.

With this image of the world, and the few equations that make it concrete, we can describe almost everything that we see.

Almost. Something is missing. And it is this something that we are seeking. The rest of the book talks about this missing part.

Turning the page, you pass from what, for good or ill, we credibly know about the world, to what we don't yet know but are trying to glimpse.

Turning the page is like leaving the security of our small spacecraft of near-certainties and stepping into the unknown.

5. Spacetime is Quantum

There is a paradox at the heart of our understanding of the physical world. General relativity and quantum mechanics, the two jewels that the twentieth century has left us, have been prolific in gifts – for comprehending the world and for today's technology. From the first of these, cosmology has developed, as well as astrophysics, the study of gravitational waves and of black holes. The second has provided the foundation for atomic physics, nuclear physics, the physics of elementary particles and of condensed matter, and of much else besides.

And yet between the two theories there is something that grates. They cannot both be true, at least not in their present forms, because they appear to contradict each other. The gravitational field is described without taking quantum mechanics into account, without accounting for the fact that fields are quantum fields – and quantum mechanics is formulated without taking into account the fact that spacetime curves and is described by Einstein's equations.

A university student attending lectures on general relativity in the morning, and others on quantum mechanics in the afternoon, might be forgiven for concluding that his professors are fools, or that they haven't talked to each other for at least a century. In the morning, the world is a *curved* spacetime where everything is *continuous*; in the afternoon, the world is a *flat* one where *discrete* quanta of energy leap and interact.

The paradox resides in the fact that both theories work remarkably well.

With every experiment and every test, nature continues to say 'you are right' to general relativity, and continues to say 'you are right' to quantum mechanics as well, despite the seemingly opposite assumptions on which the two theories are founded. It is clear that something still eludes us.

In most situations we can neglect quantum mechanics or general relativity (or both). The Moon is too large to be sensitive to minute quantum granularity, so we can forget the quanta when describing its movements. On the other hand, an atom is too light to curve space to a significant degree, and when we describe it we can forget the curvature of space. But there are situations where both curvature of space and quantum granularity matter, and for these we do not yet have an established physical theory that works.

An example is the interior of black holes. Another is what happened to the universe during the Big Bang. In more general terms, we do not know how time and space behave at very small scale. In all these instances, today's theories become confused and no longer tell us anything reasonable: quantum mechanics cannot deal with the curvature of spacetime, and general relativity cannot account for quanta. This is the problem of quantum gravity.

The problem goes even deeper. Einstein understood that space and time are manifestations of a physical field: the gravitational field. Bohr, Heisenberg and Dirac understood that physical fields have a quantum character: granular, probabilistic, manifesting through interactions. It follows that space and time must also be quantum entities possessing these strange properties.

What, then, is quantum space? What is quantum time? This

is the problem we call quantum gravity. A band of theoretical physicists scattered across five continents is laboriously seeking to solve the problem. Their objective is to find a theory, that is to say, a set of equations – but, above all, a coherent vision of the world – with which to resolve the current schizophrenia between quanta and gravity.

It isn't the first time that physics has found itself faced with two highly successful but apparently contradictory theories. The effort to synthesize has in the past been rewarded with great strides forward in our understanding of the world. Newton discovered universal gravity precisely by combining Galileo's physics of how things move on Earth with Kepler's physics of the heavens. Maxwell and Faraday found the equations of electromagnetism by bringing together what was known about electricity and what was known about magnetism. Einstein found special relativity in order to resolve the apparent conflict between Newton's mechanics and Maxwell's electromagnetism – and then general relativity in order to resolve the resulting conflict between Newton's mechanics and his own special relativity.

Theoretical physicists are thus only too happy when they discover a conflict of this type: it is an extraordinary opportunity. The question to ask is: can we construct a conceptual structure compatible with what we have learned about the world with *both* theories?

To comprehend what quantum space and quantum time are, we need once more to revise in depth the way we conceive things. We need to rethink the grammar of our understanding of the world. Just as happened with Anaximander, who understood that Earth flies in space, and that 'up' and 'down' do not exist in the cosmos; or with Copernicus, who understood that we are moving across the heavens

Figure 5.1 Matvei
Bronštejn.

at great speed; or with Einstein, who understood that spacetime squashes like a mollusc, and time passes differently in different places . . . once again, in seeking a coherent vision of the world in keeping with what we have learned about it, our ideas about the nature of reality have to change.

The first to realize that our conceptual basis must change in order to understand quantum gravity was a romantic and legendary character: Matvei Bronštejn, a young Russian who lived during the Stalin era and died tragically.

Matvei

Matvei was a younger friend of Lev Landau – the scientist who would go on to become the best theoretical physicist of the Soviet Union. Colleagues who knew them both would claim that, of the two, Matvei was the more brilliant. At the point when Heisenberg and Dirac were constructing the bases of quantum mechanics, Landau, wrongly, thought that fields became ill defined due to quanta: quantum fluctuation would prevent us from measuring the value of a component of a field at a point (an arbitrary small region) in space. The masterly Bohr immediately saw that Landau was wrong, studied the issue in depth and wrote a long and detailed article to show that fields, such as the electric one, remain well defined even when quantum mechanics is brought to bear.[1] Landau dropped the issue.

But his young friend Matvei was intrigued, realizing that Landau's intuition, though imprecise, contained something of importance. He repeated the same reasoning by which Bohr had demonstrated that the quantum electric field was well defined at a point of space, applying it instead to the gravitational field, for which Einstein had just a few years previously written the equations. And here – surprise! – Landau was right. The gravitational field at a point is *not* well defined, when taking quanta into account.

There is an intuitive way of understanding what happens. Suppose we want to observe a very, very, very small region of space. To do this, we need to place something in this area, to mark the point that we wish to consider. Say we place a particle there. Heisenberg had understood that you can't locate a particle at a point in space for long. It soon escapes. The smaller the region in which we try to locate a particle, the greater the velocity at which it escapes. (This is Heisenberg's uncertainty principle.) If the particle escapes at great speed, it has a great deal of energy. Now let us take Einstein's theory into account. Energy makes space curve. A lot of energy means that space will curve a great deal. A *lot* of energy in a *small* region results in curving space so much that it collapses into a black hole, like a collapsing star. But if a particle plummets into a black hole, I can no longer see it. I can no longer use it as a reference point for a region of space. I can't manage to measure arbitrarily small regions of space, because if I try to do this these regions disappear inside a black hole.

This argument can be made more precise with a little mathematics. The result is general: quantum mechanics and general relativity, taken together, imply that there is a limit to the divisibility of space. Below a certain scale, nothing more is accessible. More precisely, nothing exists there.

How small is this minimal region of space? The calculation is easy: we need only to calculate the minimum size of a particle before it falls into its own black hole, and the result is straightforward. The minimum length is around:

$$L_P = \sqrt{\frac{\hbar G}{c^3}}$$

Under the sign of the square root there are the three constants of nature we have already encountered: Newton's constant G, discussed in Chapter 2, which sets the strength of gravity; the speed of light c, introduced in Chapter 3 when discussing relativity, which opens up the extended present; and Planck's constant $\hbar$, found in Chapter 4, which determines the scale of the quantum granularity.* The presence of these three constants confirms the fact that we are looking at something which has to do with gravity (G), relativity (c) and quantum mechanics ($\hbar$).

The length L_P, determined in this fashion, is called the Planck length. It should be called the Bronštejn length, but such is the way of the world. In numerical terms, it is equivalent to approximately one millionth of a billionth of a billionth of a billionth of a centimetre (10^{-33} centimetres). So, that is to say . . . small.

It is at this extremely minute scale that quantum gravity

---

* The mark on the $\hbar$ of Planck's constant serves only to indicate that Planck's constant is in this equation divided by 2π, a rather useless and idiosyncratic addition by theoretical physicists: placing the small, hard, angular mark on the $\hbar$ 'makes it elegant'.

manifests itself. To give an idea of the smallness of the scale we are discussing: if we enlarged a walnut shell until it had become as big as the whole observable universe, we would still not see the Planck length. Even after having been enormously magnified thus, it would still be a million times smaller than the actual walnut shell was before magnification. At this scale, space and time change their nature. They become something different; they become 'quantum space and time', and understanding what this means is the problem.

Matvei Bronštejn understands all of this in the 1930s and writes two short and illuminating articles in which he points out that quantum mechanics and general relativity, taken together, are incompatible with our customary idea of space as an infinitely divisible continuum.[2]

There is, however, a problem. Matvei and Lev are sincere communists. They believe in revolution as the liberation of mankind, the construction of a genuinely better society, without injustice, without the immense inequalities which we still see growing systematically throughout the world. They are enthusiastic followers of Lenin. When Stalin assumes power, they are both perplexed, then critical, then hostile. They write articles which are mildly but openly critical . . . This was not the communism they wanted . . .

But these are harsh times. Landau gets through them, not easily, but he survives. Matvei, the year after having been the first to understand that our ideas on space and time had to change in a radical way, is arrested by Stalin's police and condemned to death. His execution takes place on the same day as his trial, 18 February, 1938.[3] He is thirty years old.

John

After Matvei Bronštejn's premature death, many of the century's eminent physicists tried to solve the puzzle of quantum gravity. Dirac dedicated the final years of his life to the problem, opening avenues and introducing ideas and techniques on which a good part of current work on quantum gravity is based. It is thanks to these techniques that we know how to describe a world without time, as I will explain further on. Feynman tried, attempting to adapt the techniques he had developed for electrons and photons to the context of general relativity, but without success: electrons and photons are quanta in space; quantum gravity is something else: it isn't enough to describe 'gravitons' moving in space, it is space itself that has to be quantized.

A few Nobel Prizes were awarded to physicists who happened to resolve other problems, almost by mistake, during the course of their attempts to disentangle the puzzle of quantum gravity. Two Dutch physicists, Gerard 't Hooft and Martinus Veltman, received the Nobel Prize in 1999 for having shown the consistency of the theories which today are used to describe nuclear forces – a part of the standard model – but their research programme was actually aiming to demonstrate the consistency of a theory of quantum gravity. They were working on the theories of these other forces only as a preliminary exercise. The 'preliminary exercise' earned them a Nobel Prize, but they did not succeed in showing the consistency of their version of quantum gravity.

The list could go on and would read like a roll of honour of the century's outstanding theoretical physicists. As well as like a catalogue of failures. Very gradually, though, over the course of decades, ideas were clarified and dead ends explored and usefully closed off; techniques and general ideas were strengthened, and results began to build, one developing from another. To mention here the numerous scientists who have contributed to this gradual, slow-moving, collective construction would require a tedious list of names, each one of whom has added a grain or a stone to the process.

I would like to mention just one, who for years held together the threads of this collective research: the remarkable, eternally youthful Englishman – half philosopher and half physicist – Chris Isham. It was when reading one of his articles reviewing the question of quantum gravity that I first became enamoured with the problem. The article explained just why it was so difficult, how our conception of space and time needed to be modified, and gave a lucid overview of all the routes which were being followed at the time, with the results achieved, and difficulties entailed. I was in my third year at university, and the possibility of rethinking space and time from square one fascinated me. This fascination has never diminished.
For, as Petrarch sings, 'The wound does not heal due to the weakening of the bow.'

The scientist who has most contributed to quantum gravity is John Wheeler, a legendary figure who has traversed the physics of the past century. A pupil of and collaborator with Niels

Figure 5.2 John Wheeler.

Bohr in Copenhagen; a collaborator with Einstein when Einstein moved to the United States; a teacher who can count among his students figures such as Richard Feynman ... Wheeler was at the heart of the physics of the twentieth century. He was gifted with a fervid imagination. It was he who invented and made popular the term 'black hole'. His name is associated with the early extended investigations – frequently more intuitive than mathematical – into how to think about quantum spacetime. Having absorbed Bronštejn's lesson that quantum properties of the gravitational field imply a modification of the notion of space at a small scale, Wheeler looked for novel ideas to help conceive of this quantum space. He imagined it as a cloud of superimposed geometries, just as we can think of a quantum electron as a cloud of positions.

Imagine that you are looking at the sea from a great height: you perceive a vast expanse of it, a flat, cerulean table. Now you descend and look at it more closely. You begin to make out the great waves swollen by the wind. You descend further, and you see that the waves break up and that the surface of the sea is a turbulent frothing. This is what space is like, as imagined by Wheeler.* On our scale, immensely larger than the Planck length, space is smooth. If we move down to the Planck scale, it shatters and foams.

Wheeler sought a way to describe this foaming of space, this wave of probability of different geometries. In 1966, a young colleague of his who lived in Carolina, Bryce DeWitt, provided the key.[4] Wheeler travelled frequently, and met collaborators wherever he could. He asks Bryce to meet at

* To hear this metaphor directly in his own voice, go to the site http://www.webofstories.com/play/9542?o=MS.

Raleigh Durham airport, in North Carolina, where he had a few hours' wait between connecting flights. Bryce arrives and shows him an equation for 'a wave function of space', obtained by using a simple mathematical trick.* Wheeler is enthused. From this conversation a type of 'equation of orbitals' for general relativity is born; an equation which should determine the probability of one or another curved spaces. For a long time, DeWitt called it Wheeler's equation† – while Wheeler called it the DeWitt equation. Everyone else calls it the Wheeler-DeWitt equation.

The idea is very good, and becomes a basis for the attempts to construct the full theory of quantum gravity. But the equation itself is riddled with problems – serious ones. In the first place, from a mathematical point of view, the equation is really quite badly defined. If we try to use it to do calculations, we soon obtain results that are infinite, which makes no sense. It must be improved.

But it is also difficult to understand how to interpret this equation, to know what it means. Among its disconcerting aspects is the fact that it no longer contains the time variable. How can it be used to compute the evolution of something which happens in time if it does not include a time variable? Dynamical equations, in physics, always contain the variable t, time. What does a physical theory without a temporal variable signify? For years, research will revolve around such

* DeWitt replaces derivatives with derivative operators in the Hamilton–Jacobi equation for general relativity (written a little while earlier by Peres). That is, he does what Schrödinger had done to write his equation, in his first work: replacing derivatives with derivative operators in the Hamilton–Jacobi equation of a particle.
† Or the 'Einstein–Schrödinger' equation.

questions, trying to revise the equation in different manners, in order to improve its definition and understand what it might mean.

The first steps of the loops

The fog begins to dissipate towards the end of the 1980s. Surprisingly, some solutions of the Wheeler–DeWitt equation appear. During these years I found myself first at the University of Syracuse, in New York State, visiting the Indian physicist Abhay Ashtekar, and then in Connecticut, at Yale University, visiting the American physicist Lee Smolin. I remember a period of intense discussions and burning intellectual fervour. Ashtekar had rewritten the Wheeler–De Witt equation in a simpler form; and Smolin, together with Ted Jacobson of the University of Maryland in Washington, had been the first to find some of these new strange solutions of the equation.

The solutions had a curious peculiarity: they depended on *closed lines* in space. A closed line is a 'loop'. Smolin and Jacobson could write a solution to the Wheeler–DeWitt equation for every loop: for every line closed on itself. What did this mean? The first works of what will later become known as loop quantum gravity emerge from these discussions, as the meaning of these solutions of the Wheeler–DeWitt equation gradually clarify. Upon these solutions, little by little, a coherent theory begins to be erected, inheriting the name 'loop theory' from the first solutions studied.

Today there are hundreds of scientists working on this theory, spread throughout the world from China to Argentina, from Indonesia to the United States. What is slowly being erected is the theory now known as loop theory, or loop quantum gravity: the theory to which the following chapters are devoted. It is not the only direction explored in the search for a quantum theory of gravity, but it is the one I consider the most promising.*

* The best known alternative to loop quantum gravity is string theory, whose main concern is not so much studying the quantum properties of space and time, but rather writing a unified theory of all known fields, an objective that might be premature given current knowledge.

6. Quanta of Space

The last chapter closed with the solutions of the Wheeler–DeWitt equation discovered by Jacobson and Smolin. These solutions depend on lines that close on themselves, or loop. What does it all mean?

Remember Faraday's lines – the lines which carry the electric force and which, in Faraday's vision, fill space? The lines from which the concept of 'field' originates? Well, the closed lines that appear in the solutions of the Wheeler–DeWitt equation are Faraday lines of the gravitational field.

But two new ingredients are now added to Faraday's ideas.

The first is that we are dealing with quantum theory. In quantum theory, everything is discrete. This implies that the infinitely fine, continuous spiderweb of Faraday's lines now becomes similar to a real spiderweb: it has a finite number of distinct threads. Every single line determining a solution of the Wheeler–DeWitt equation describes one of the threads of this web.

The second new aspect, the crucial one, is that we are speaking of gravity and, therefore, as Einstein understood, we are not speaking of fields immersed in space but of the very structure of space itself. Faraday's lines of the quantum gravitational field are the threads of which space is woven.

At first, the research was focused on these lines and how they could 'weave' our three-dimensional physical space. Figure 6.1 represents an early attempt to give an intuitive idea of the discrete structure of space which would result from this.

Soon after, thanks to the intuitions and mathematical ability of young scientists such as the Argentine Jorge Pullin and the Pole Jurek Lewandowski, it became clear that the key to understanding the physics of these solutions lies in the points where these lines intersect. These points are called nodes, and the lines between nodes are called links. A set of intersecting lines forms what is called a graph, that is to say, a combination of nodes connected by links, as in figure 6.3.

A calculation, in fact, demonstrates that, without nodes, physical space has no volume. In other words, it is in the nodes of the graph, not in the lines, that the volume of space 'resides'. The lines 'link together' individual volumes sitting at the nodes.

Getting to a full clarification of the resulting picture of quantum spacetime took years. It was necessary to transform the ill-defined mathematics of the Wheeler–De Witt equation into a structure sufficiently well defined to be able to compute with. With this, it became possible to achieve

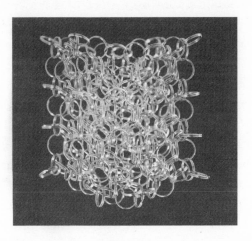

Figure 6.1 The quantum version of Faraday's lines of force, which weave space like a three-dimensional mesh of interlinked rings (loops).

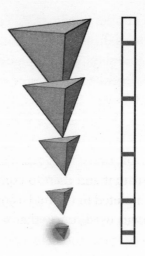

Figure 6.2 The spectrum of the volume: the volumes of a regular tetrahedron that are possible in nature are limited in number. The smallest, at the bottom, is the smallest volume in existence.

precise results. The key technical result which clarifies the physical meaning of the graphs is the calculation of the spectra of volume and of area.

Spectra of volume and area

Take any region of space, for example, the room in which you are reading this, if you are in a room. How big is this room? The size of the space of the room is measured by its volume. Volume is a geometrical quantity which depends on the geometry of space, but the geometry of space – as Einstein understood, and as I recounted in Chapter 3 – is the gravitational field. Volume is therefore a property of the gravitational field, expressing how much gravitational field there is between the walls of the room. But the gravitational field is a physical quantity and, like all physical quantities, is subject to the laws of quantum mechanics. In particular, like all physical quantities, volume may not assume arbitrary values but only certain particular ones, as I described in Chapter 4. The list of all possible values is called, if you remember, the spectrum. Hence there should exist a 'spectrum of the volume' (figure 6.2)

Dirac provided us with the formula with which to compute the spectrum of every variable. The calculation took time, first to formulate it and then to complete it, and made us suffer. It was completed in the mid-1990s, and the answer, as expected (Feynman used to say that we should never do a calculation without first knowing the result), is that the spectrum of the volume is discrete. That is, the volume can only be made up of 'discrete packets'. These are somewhat similar to the energy of the electromagnetic field, which is also formed of discrete packets: photons.

The nodes of the graph represent the discrete packets of

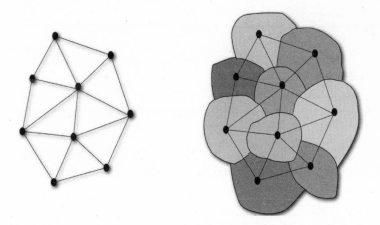

Figure 6.3 On the left, a graph formed by nodes connected by links. On the right, the grains of space which the graph represents. The links indicate the adjacent particles, separated by surfaces.

volume and, as in the case of photons, can only have certain sizes, which can be computed using Dirac's general quantum equation.* Every node n in the graph has its own volume v_n: one of the numbers in the spectrum of the volume. The nodes are the elementary quanta of which physical space is made. Every node of the graph is a 'quantum particle of space'. The structure that emerges is the one illustrated in figure 6.3.

A link is an individual quantum of a Faraday line. Now we can understand what it represents: if you imagine two nodes as two small 'regions of space', these two regions will be separated by a small surface. The size of this surface is its *area*. The second quantity, after the volume, which characterizes the quantum webs of space, is the area associated with each line.†

The area, just as in the case of the volume, is a physical variable, and has a spectrum which may be calculated using Dirac's equation.‡ Area is not continuous, it is granular. There is no such thing as an arbitrarily small area.

* The eigenvalue equation for the volume operator.
† Hence the quantum states of gravity are indicated with $| j_l, v_n >$, where n indicates the nodes and l the links of the graph.
‡ The result of the calculation is simple. I show it here so you can see how Dirac's spectra work. The possible values of the area A are given in the following formula, where j is a 'half-integer', that is to say, a number which is half of an integer, such as 0, ½, 1, ½, 2, ½, 3 . . .

$$A = 8\pi L_p^2 \sqrt{j(j+1)}$$

A is the area that a surface separating two grains of space can have. 8 is the number eight, nothing special about it. π is the Greek pi which we studied at school: the constant which gives the relation between the circumference and the diameter of any circle, and which appears everywhere in physics, I don't know why. L_p is the Planck length, the extremely

Space appears continuous to us only because we cannot perceive the extremely small scale of these individual quanta of space. Just as when we look closely at the cloth of a T-shirt, we see that it is woven from small threads.

---

small scale at which the phenomena of quantum gravity take place. L^2_p is the square of L_p, which is the (extremely small) area of a tiny square with sides equal to the Planck length. Therefore $8\pi L^2_p$ is simply a 'small' area: the area of a minuscule square with a side which is about a millionth of a billionth of a billionth of a billionth of a centimetre (10^{-66} cm²). The interesting aspect of the formula is the square root and what is within it. The key point is that j is a half-integer, that is to say, it may have only values which are multiples of ½. For each one of these, the root has a certain value, listed approximately in table 6.1.

Table 6.1 Spin (half-integers) and corresponding value of the area in units of minimal area.

j	$\sqrt{j(j+1)}$
½	0.8
1	1.4
3/2	1.9
2	2.4
5/2	2.9
3	3.4
—	—

Multiplying the numbers in the right-hand column by the area $8\pi L^2_p$, we obtain the possible values of the area of the surface. These special values are like the ones which appear in the study of the orbits of electrons in atoms, where quantum mechanics allows only certain orbits. The point is that no other areas apart from the values derived by this equation *exist*. No surface can have an area one tenth of $8\pi L^2_p$.

When we say that the volume of a room is, for example, 100 cubic metres, we are in effect *counting* the grains of space – the quanta of the gravitational field – which it contains. In a room, this number has more than a hundred digits. When we say that the area of this page is 200 square centimetres, we are actually counting the number of links in the web, or loops, which traverse the page. Across the page of this book, there is a number of quanta with more or less seventy digits.

The idea that measuring length, area and volume is a question of counting individual elements had been proposed in the nineteenth century by Riemann himself. The mathematician who had developed the theory of *continuous* curved mathematical spaces was already aware that a *discrete* physical space is, ultimately, more reasonable than a continuous one.

To summarize, the theory of loop quantum gravity, or loop theory, combines general relativity with quantum mechanics in a rather conservative way, because it does not employ any other hypothesis apart from those of the two theories themselves, suitably rewritten to render them compatible. But the consequences are radical.

General relativity taught us that space is something dynamic, like the electromagnetic field: an immense, mobile mollusc in which we are immersed, which stretches and bends. Quantum mechanics teaches us that every field of this sort is made of quanta, that is to say, it has a fine, granular structure. It follows that physical space, being a field, is made of quanta as well. The same granular structure characterizing the other quantum fields also characterizes the quantum gravitational field, and therefore space. We expect space to be granular. We expect quanta of gravity, just as there are quanta of light, quanta of the electromagnetic field, and as particles are quanta of quantum fields. But space is the gravitational field, and the quanta of the gravitational field are quanta of space: the granular constituents of space.

The central prediction of loop theory is therefore that space is not a continuum, it is not divisible ad infinitum, it is formed of 'atoms of space'. A billion billion times smaller than the smallest of atomic nuclei.

Loop theory describes this atomic and granular quantum structure of space in a precise mathematical form. It is obtained by applying the general equations of quantum mechanics written by Dirac to Einstein's gravitational field.

In particular, loop theory specifies that volume (for example, the volume of a given cube) cannot be arbitrarily small. A minimum volume exists. No space smaller than this minimum volume exists. There is a minimum quantum of volume: an elementary atom of space.

Atoms of space

Remember Achilles chasing after the tortoise? Zeno observed that there is something difficult to accept in the idea that Achilles has to cover an infinite number of distances before reaching the slow-moving creature. Mathematics had found a possible answer to this difficulty, showing how an infinite number of progressively smaller intervals could nevertheless amount to a finite total interval.

But is this what *truly* happens in nature? Are there intervals between Achilles and the tortoise that can be arbitrarily short? Does it really make sense to talk of a billionth of a billionth of a billionth of a millimetre, and then to think of dividing it again further innumerable times?

The calculation of the quantum spectra of geometric quantities indicates that the answer is negative: arbitrarily small chunks of space do not exist. There is a lower limit to the divisibility of space. It is at a very small scale indeed, but it is there. This is what Matvei Bronštejn had intuited in the 1930s. The calculation of the spectra of volume and area confirms Bronštejn's idea and frames it in a mathematically precise manner.

Achilles does not need an infinite number of steps to reach the tortoise because, in a space made of grains of finite size, infinitely small steps do not exist. The hero will come ever closer to the creature until, in the end, he reaches it in a single quantum leap.

But, on reflection, was this not precisely the solution proposed by Leucippus and by Democritus? They spoke of the granular structure of matter, and we are rather unsure as to what, precisely, they said about space. Unfortunately, we do not have their texts and must make do with the sparse fragments in the citations of others. It is like trying to reconstruct Shakespeare's plays from a list of Shakespeare quotes.* Democritus's argument on the incongruity of the continuum as a collection of points, reported by Aristotle, may be applied to space. I imagine that if we could ask Democritus if it makes sense to split a space interval ad infinitum, his reply could only be to repeat that divisibility must have a limit. For the philosopher of Abdera, matter is made of atoms that cannot be divided. Having once understood that

---

* Imagine what a nonsensical hotchpotch the ideas of Aristotle and Plato would seem if we only had the commentaries on them written by others and were unable to access the lucidity and complexity of the original texts!

space is very much like matter – space, as he had said himself, has its own nature, 'a certain physics' – I suspect he would not have hesitated to deduce that space, too, can only be made of elementary chunks that cannot be divided. We are perhaps just following in the footsteps of Democritus.

I certainly don't mean to imply that the physics of two millennia was useless, that experiments and mathematics are pointless and that Democritus could be as convincing as modern science. Obviously not. Without experiments and mathematics, we would never have understood what we have understood. Yet we develop our conceptual schema for understanding the world by exploring new ideas but also by building on the powerful intuitions of giant figures from the past. Democritus is one of them, and we discover the new sitting on his titanic shoulders.

But let us return to quantum gravity.

Spin networks

The graphs which describe the quantum states of space are characterized by a volume v for every node and a half-integer j for every line. A graph with this additional information is called a spin network (figure 6.4). (Half-integers in physics are called 'spin' because they appear in the quantum mechanics of spinning objects.) A spin network represents a quantum state of the gravitational field: a quantum state of space; a granular space in which area and volume are discrete. Fine-mesh grids are used elsewhere in physics to approximate continuous space. Here, there is no space continuum to approximate: space is genuinely granular.

The crucial difference between photons (the quanta of the

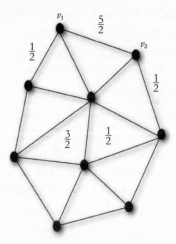

Figure 6.4 A spin network

electromagnetic field) and the nodes of the graph (the quanta of gravity) is that photons exist in space, whereas the quanta of gravity constitute space themselves. Photons are characterized by 'where they are'.* Quanta of space have no place to be in, because they are themselves that place. They have only one piece of information which characterizes them spatially: information about which other quanta of space they are adjacent to, which one is next to which other. This information is expressed by the links in the graph. Two nodes connected by a link are two nodes in proximity. They are two grains of space in contact with each other: this 'touching' constructs the structure of space.

The quanta of gravity, that is, are not *in* space, *they are themselves space*. The spin networks which describe the quantum

* The quantum number of the states of photons in Fock's space is the momentum, Fourier's transformation of position.

structure of the gravitational field are not immersed in space; they do not inhabit a space. The location of single quanta of space is not defined with regard to something else but only by the links and the relation these express.

If I step from grain to grain along the links until I complete a circuit and return to the grain from which I started, I will have made a 'loop'. These are the original loops of the loop theory. In Chapter 4 I showed that the curvature of space may be measured by looking at whether an arrow transported across a closed circuit returns pointing in the same direction, or turned. The mathematics of the theory determines this curvature for every closed circuit on a spin network, and this makes it possible to evaluate the curvature of spacetime, and hence the force of the gravitational field, from the structure of a spin network.*

Now, quantum mechanics is more than granularity. There is also the fact that evolution is probabilistic – the way in which the spin networks evolve is random. I'll speak about this in the next chapter, devoted to time.

And there is the fact that what matters is not how things *are*, but rather how they *interact*. Spin networks are not entities; they describe the effect of space upon things. Just as an electron is in no place but diffused in a cloud of probability in all places, space is not actually formed by a single specific spin network but rather by a cloud of probabilities over the whole range of all possible spin networks.

At an extremely small scale, space is a fluctuating swarm of quanta of gravity which act upon each other, and together

---

* The operator associated with the geometry of granular space is the holonomy of the gravitational connections, or rather, in physical terms, a 'Wilson loop' of general relativity.

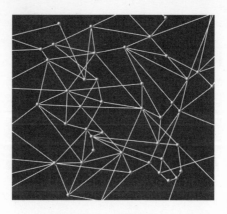

Figure 6.5 At a minute scale, space is not continuous:
it is woven from interconnected finite elements.

act upon things, manifesting themselves in these interactions as spin networks, grains interrelated with each other (figure 6.5).

Physical space is the fabric resulting from the ceaseless swarming of this web of relations. The lines themselves are nowhere; they are not in a place but rather create places through their interactions. Space is created by the interaction of individual quanta of gravity.

This is the first step towards understanding quantum gravity. The second concerns time. And, to time, the next chapter is devoted.

7. Time Does Not Exist

It must not be claimed that anyone can sense time by itself
apart from the movement of things.

– Lucretius, *De rerum natura*[1]

The alert reader will have realized that in the preceding
chapter little attention was given to time. And yet Einstein
showed, over a century ago, that we cannot separate time and
space, that we must think of them together as a single whole:
spacetime. The moment has come to rectify this and bring
time back into the picture.

Research on quantum gravity has revolved for years
around spatial equations, before having the courage to con-
front time. In the last fifteen years, a way of thinking about
time has begun to emerge. I'll try to explain it.

Space as an amorphous container of things disappears
from physics with quantum gravity. Things (the quanta) do
not inhabit space, they dwell one over the other, and space is
the fabric of their neighbouring relations. As we abandon
the idea of space as an inert container, similarly, we must
abandon the idea of time as an inert flow along which reality
unfurls. Just as the idea of the space continuum containing
things disappears, so, too, does the idea of a flowing continuum 'time' during the course of which phenomena happen.

In a certain sense, space no longer exists in fundamental

theory; the quanta of the gravitational field are not *in* space. In the same sense, time no longer exists in the fundamental theory: the quanta of gravity do not evolve *in* time. Time just counts their interactions. As evidenced with the Wheeler–De Witt equation, the fundamental equations no longer contain the time variable. Time emerges, like space, from the quantum gravitational field.

This was already partially true for classical general relativity, where time already appears as an aspect of the gravitational field. But as long as we neglect quantum theory, we can still think of spacetime in a rather conventional manner, like the tapestry in which the story of the rest of reality unfolds, even if it is a dynamical, moving tapestry. The moment we take quantum mechanics into account, we recognize that time, too, must have those aspects of probabilistic indeterminacy, granularity and relationality which are common to all of reality. It becomes a 'time' markedly different from all that we have hitherto meant by the word.

This second conceptual consequence of the theory of quantum gravity is more extreme even than the vanishing of space.

Let's attempt to understand it.

Time is not what we think it is

That the nature of time is different from the common idea which we have of it was already clear over a century ago. Special and general relativity made this explicit. Today, the inadequacy of our common-sense view of time can be easily verified in a laboratory.

Let's reconsider, for example, the first consequence of

general relativity, as illustrated in Chapter 3. Take two watches, ensure that they mark exactly the same time, place one on the floor and the other on a piece of furniture. Wait for about half an hour and then bring them back next to each other. Will they still tell the same time?

As described in Chapter 3, the answer is no. The watches which we usually wear on our wrists, or have on our mobile phones, are not sufficiently precise to allow us to verify this fact, but in physics laboratories all over the world there are timepieces precise enough to demonstrate the discrepancy which occurs: the watch left on the floor is slow when compared to the one which has been raised above it.

Why? Because time does not pass in the same way everywhere in the world. In some places, it flows more quickly; in others, more slowly. The closer you get to the Earth, where gravity* is more intense, the slower time passes. Remember the twins in Chapter 3, who ended up with different ages as a result of having lived one at sea level and one in the mountains? The effect is very slight: the time gained during a life spent by the sea, with respect to one passed in the mountains, consists of fractions of a second – but the smallness of the amount does not alter the fact that there is a real difference. Time does not work as we customarily imagine it does.

We must not think of time as if there were a great cosmic clock that marks the life of the universe. We have known for more than a century that we must think of time instead as a localized phenomenon: every object in the universe has its own time running, at a pace determined by the local gravitational field.

* The gravitational potential.

But even this notion of a localized time no longer works when we take the quantum nature of the gravitational field into account. Quantum events are no longer ordered by the passage of time at the Planck scale. Time, in a sense, ceases to exist.

What does it mean to say that time does not exist?

First, the absence of the variable time from the fundamental equations does not imply that everything is immobile and that change does not happen. On the contrary, it means that change is ubiquitous. Only: elementary processes cannot be ordered along a common succession of instants. At the extremely small scale of the quanta of space, the dance of nature does not develop to the rhythm kept by the baton of a single orchestral conductor: every process dances independently with its neighbours, following its own rhythm. The passing of time is intrinsic to the world, it is born of the world itself, out of the relations between quantum events which *are* the world and which themselves generate their own time.

In fact, the nonexistence of time does not mean anything particularly complicated. Let's try to understand.

The candle chandelier and the pulse

Time appears in most equations of classic physics. It is the variable indicated by the letter t. The equations tell us how things change *in time*. If we know what has happened in the past, they allow us to predict the future. More precisely, we measure some variables – for example, the position A of an object, the angle B of a swinging pendulum, the temperature C of an object – and the equations of physics tell us how these variables *A*, *B* and *C* will change with time. They predict the functions $A(t)$, $B(t)$, $C(t)$, and so on, which describe the changing of these variables in time *t*.

Galileo was the first to understand that the movement of objects on Earth could be described by equations for the functions of time $A(t)$, $B(t)$, $C(t)$ – and the first to write explicit equations for these functions. The first law of terrestrial physics found by Galileo, for example, describes how an object falls, that is to say, how its altitude *x* varies with the passage of time *t*.*

To discover and verify this law, Galileo needed two kinds of measurements. He had to measure the height *x* of the object and the time *t*. Therefore, he needed, in particular, an instrument to measure time. He needed a *clock*.

When Galileo lived there were no accurate clocks. Galileo himself, as a young man, discovered a key to making precise timepieces. He discovered that the oscillations of a pendulum all have the same duration (irrespective of the amplitude). Thus, it is possible to measure time by simply counting the oscillations of a pendulum. It seems such an obvious idea, but it took Galileo to find it; it had not occurred to anyone before him. So it goes, with science.

But things are not really this straightforward.

According to legend, Galileo alighted on the idea in Pisa's marvellous cathedral while watching the slow oscillations of a gigantic candle chandelier, which is still there. (The legend is false, since the chandelier was actually first hung there years after Galileo's death, but it makes for a good story. Perhaps there was another one hanging there at the time.) The scientist was observing the oscillations during a religious service in

* $x(t) = \frac{1}{2} at^2$.

which he was evidently not particularly absorbed, and he was measuring the duration of each oscillation of the chandelier by counting the beats of his own pulse. With mounting excitement, he discovered that the number of beats was the same for each oscillation: it did not change when the chandelier slowed and oscillated with diminished amplitude. The oscillations all had the same duration.

It's a fine story but, on reflection, it leaves us perplexed – and this perplexity goes to the heart of the problem of time. How could Galileo know that *his own individual pulse-beats* all lasted for the same amount of time?*

Not many years after Galileo, doctors began to measure their patients' pulses by using a watch – which is nothing, after all, but a pendulum. So we use the beats to assure ourselves that the pendulum is regular, and then the pendulum to ascertain the regularity of the pulse-beats. Is this not somewhat circular? What does it mean?

It means that we, in reality, never measure time itself; we always measure the physical variables A, B, C . . . (oscillations, beats, and many other things) and compare one variable with another, that is to say, we measure the functions $A(B)$, $B(C)$, $C(A)$, and so on. We can count how many beats for each oscillation; how many oscillations for every tick of my stopwatch; how many ticks of my stopwatch between intervals of the clock on the bell-tower . . .

The point is that it is *useful* to imagine that a variable t exists – the 'true time' – which underpins all those movements, even if we cannot measure it directly. We write the equations for the physical variables with regard to this *unobservable t*, equations which tell us how things change in t; that

* Especially since he had become excited . . .

is, for instance, how much time it takes for each oscillation, and how long each heart-beat lasts. From this, we can derive how the variables change in relation to each other – how many heartbeats there are in one oscillation – and compare this prediction with what we observe in the world. If the predictions are correct, we trust that this complicated schema is a sound one and, in particular, that it is useful to employ the variable of time t, even if we cannot measure it directly.

In other words, the existence of the variable time is a useful assumption, not the result of an observation.

It was Newton who understood all of this: he understood that this was a good way to proceed, and clarified and developed this schema. Newton asserts explicitly in his book that we can't ever measure the true time t but, if we *assume* that it exists, we can set up an efficient framework to describe nature.

Having clarified this, we can return to quantum gravity and the meaning of the statement that 'time does not exist'. It simply means that the Newtonian schema no longer works when we are dealing with small things. It was a good one, but only for large things.

If we want to understand the world widely, if we want to understand how it functions in the less familiar situations where quantum gravity matters, we need to abandon this schema. The idea of a time t which flows by itself, and in relation to which all things evolve, is no longer a useful one. The world is not described by equations of evolution in time t. What we must do is simply to enumerate the variables $A, B, C \dots$ which we *actually* observe, and write equations expressing relations between *these* variables, and nothing else: that is, equations for the relations $A(B), B(C), C(A) \dots$ which we observe, and not for the functions $A(t), B(t), C(t) \dots$ which we *do not* observe.

In the example of the pulse and the candle chandelier, we

will not have the pulse and the candelabrum evolving in time, but only equations which tell us how the two variables evolve with respect to each other. That is to say, equations which tell us directly how many pulse-beats there are in an oscillation, without mentioning t.

'Physics without time' is physics in which we speak only of the pulse and the chandelier, without mentioning time.

It's a simple change – but from a conceptual point of view, it's a huge leap. We must learn to think of the world not as something which changes in time but in some other way. Things change only in relation to one another. At a fundamental level, there is no time. Our sense of the common passage of time is only an approximation which is valid for our macroscopic scale. It derives from the fact that we perceive the world in a coarse-grained fashion.

The world described by the theory is thus far from the one we are familiar with. There is no longer space which *contains* the world, and no longer time *during the course of which* events occur. There are elementary processes in which the quanta of space and matter continuously interact with each other. Just as a calm and clear Alpine lake is made up of a rapid dance of a myriad of minuscule water molecules, the illusion of being surrounded by continuous space and time is the product of a long-sighted vision of a dense swarming of elementary processes.

Spacetime sushi

How do these general ideas apply to quantum gravity? How can we describe change without the ideas of space as a container, or time along which the world glides?

Consider a process: for example, the collision of two billiard balls on a table's green baize. Imagine a red ball played in the direction of a yellow one; it gets close, collides, and the two balls move away in different directions. This process, like all processes, takes place in a finite zone of space – let's say on a table approximately two metres wide – and lasts for a finite interval of time – let's say three seconds. To deal with this process in the context of quantum gravity, it is necessary to include space and time in the process itself (figure 7.1).

We must not, in other words, describe only the two balls, but also all that is around them: the table and any other material objects – and the space in which they are immersed during the time that elapses between the start of the shot and the end of the process. Space and time are the gravitational field, Einstein's 'mollusc': we are also including the gravitational

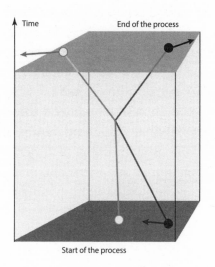

Figure 7.1 A region of space in which a black ball hits a stationary white ball, propels it and rebounds. The box is the region of spacetime. Within it are drawn the trajectories of the balls.

field, that is to say, a piece of the mollusc, in the process. Everything is immersed in Einstein's great mollusc: here, imagine that you are slicing a small, finite portion of it, like a piece of sushi, which encompasses the collision and what surrounds it.

What we obtain from this is a spacetime box (as in figure 7.1): a finite portion of spacetime a few cubic metres in dimension by a few seconds of time. *This* process does not occur 'in' time. The box is not *in* spacetime, it *includes* spacetime. It isn't a process *in* time, in the same way in which grains of space are not *in* space. The passage of time is only the measure of the process itself, just as quanta of gravity are not *in* space, as they themselves constitute space.

The key to understanding how quantum gravity works lies in considering not solely the physical process given by the two balls but rather the entire process defined by the whole box with all that it entails, including the gravitational field.

Now let us return to Heisenberg's original insight: quantum mechanics does not tell us what happens during the course of a process, but the probability which ties together the different initial and final states of the process. In our case, the initial and final states are given by all that happens at the *border* of the spacetime box.

What the equations of loop quantum gravity give us is the probability associated with a given possible *boundary* of the box – the probability that the balls will come out of the box in one particular configuration or another, if they have entered it in another.

How is this probability computed? Recall Feynman's sum over paths, which I described when speaking about quantum mechanics. Probabilities, in quantum gravity, can be calculated in the same way. By considering all the possible 'trajectories' that have the same boundary. Since we are including the dynamics of spacetime, this means considering *all possible spacetimes* which have the same boundary as the box.

Quantum mechanics assumes that between the initial boundary, where the two balls enter, and the final boundary where they exit, there is no definite spacetime nor definite trajectory of the balls. There is a quantum 'cloud' in which all the possible spacetimes and all possible trajectories exist together. The probability of seeing the balls going out in one way or another can be computed by summing over all possible spacetimes.

Spinfoam

If quantum *space* has the structure of a spin network, what structure will *spacetime* have? What will one of the spacetimes previously alluded to in the calculation be like?

It must be a 'history' of a spin network. Imagine that you take the graph of the spin network and move it: every node in the web draws a line, like the balls in figure 7.1, and every line of the graph, moving, draws a surface (for example, a moving segment draws a rectangle). But there is more: a node can open up into two or more nodes, just as a particle can split into two or more particles. Conversely, two or more nodes can combine into a single one. In this way, a graph which evolves draws an image like the one in figure 7.2.

The image portrayed on the right of figure 7.2 is a 'spinfoam'. 'Foam' because it is made of surfaces which meet on

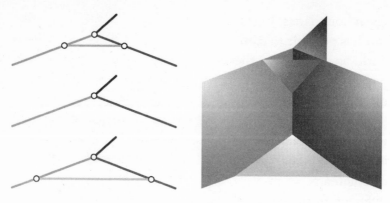

Figure 7.2 An evolving spin-network: three nodes combine into a single node, and then separate again. On the right, the spinfoam representing this process.

Figure 7.3 The foam of soap bubbles.

lines, which in turn meet on vertices, resembling a foam of soap bubbles (figure 7.3). '*Spin-foam*' because the faces of the foam carry spins, as do the links of the graphs whose evolution they describe. To compute the probability of a process, one must sum up over all the possible spinfoams within the box which have the same boundary as that process. The boundary of a spinfoam is a spin network and the matter on it.

The equations of loop quantum gravity express the probability of a process in terms of sums over spinfoams with given boundaries. In this way it is possible to compute, in principle, the probability of any physical event.*

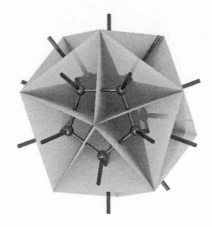

Figure 7.4 A vertex of spinfoam. Courtesy of Greg Egan.

At first sight, this way for making calculations in quantum gravity, based on spinfoams, seems very different from the usual ways in which things are computed in theoretical physics. There is no given space, no given time, and spinfoams seem objects quite remote from, say, the particles of the standard model. But in fact there are strong similarities between the spinfoam technique and the calculation techniques used in the standard model. In fact, even more than this, the spinfoam technique is actually a beautiful merging of the two main calculation techniques used in the context of the standard model: Feynman diagrams and the lattice approximation.

Feynman diagrams are used, for instance, to compute processes dominated by electromagnetic or weak forces. A Feynman diagram represents a sequence of elementary interactions among particles. An example is in figure 7.5, which represents two particles, or two quanta of the field, interacting. The particle on the left splits into two particles, one of

* The actual structure of the vertices of the spinfoam is a bit more complex than the one in figure 7.2, and resembles more closely the one shown in figure 7.4.

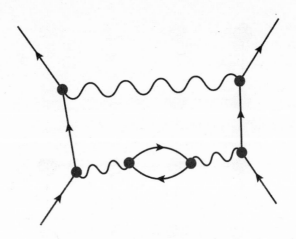

Figure 7.5 A Feynman diagram.

which splits in turn into two particles, which then reunite, converging with the particle on the right. The graph portrays a history of the field's quanta.

The lattice approximation is used when the forces are strong and the particle picture is no longer effective for describing physics, for instance in computing the strong forces between quarks inside the nucleus of an atom. The lattice technique entails approximating a continuous physical space by means of a lattice, or a grid, as in figure 7.6. This grid is not assumed to be a faithful description of space, but only an approximation, as when engineers calculate the resistance of a bridge by approximating the concrete with a finite number of elements. These two methods of making calculations – Feynman diagrams and the lattice – are the two most efficient techniques of quantum field theory.

In quantum gravity, something beautiful occurs: the two methods of making calculations become one and the same. The spacetime foam represented in figure 7.2, used to

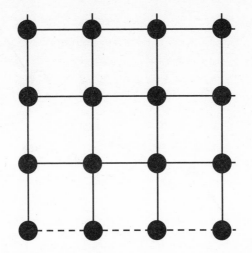

Figure 7.6 A grid approximating physical spacetime.

compute a physical process in quantum gravity, may be interpreted *either* as a Feynman diagram *or* as a lattice calculation.* Therefore, the two calculation techniques used for the standard model turn out to be particular cases of a common technique: summing over the *spinfoams* of quantum gravity.

Earlier, I set out Einstein's equations. Again, I can't resist including here the complete collection of the equations of loop theory, even if the reader will obviously not be able to decipher them – not before undertaking the study of a good

---

* It's a Feynman diagram because it is a history of quanta, as in the Feynman diagrams. Except that now, the quanta are not quanta moving in space, but rather quanta *of* space. The graph they draw in their interactions is not a representation of the movement of particles in space, but represents the plot of space itself. But the resulting picture is also precisely a lattice like the one used in the lattice approximation, because it represents a discretized spacetime. With the difference that it is no longer an approximation, but the *real* discrete structure of space at a small scale.

deal of mathematics. Someone once claimed that a theory isn't credible if its equations cannot be summarized on a T-shirt. Here is that T-shirt for loop quantum gravity (figure 7.7).

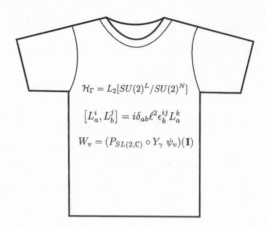

$$\mathcal{H}_\Gamma = L_2[SU(2)^L / SU(2)^N]$$

$$[L_a^i, L_b^j] = i\delta_{ab}\ell^2 \epsilon_k^{ij} L_a^k$$

$$W_v = (P_{SL(2,\mathbb{C})} \circ Y_\gamma \ \psi_v)(\mathbf{I})$$

Figure 7.7 The equations of loop quantum gravity, summarized on a T-shirt.

These equations* are the mathematical version of the picture of the world I have given in the last two chapters. We are not at all sure if they are the correct equations – but, in my opinion, they are the best account of quantum gravity we have at present.

Space is a spin network whose nodes represent its elementary grains, and whose links describe their proximity relations. Spacetime is generated by processes in which these spin networks transform into one another, and these processes are described by sums over spinfoams. A spinfoam represents a

---

* The first defines the Hilbert space of the theory. The second describes the algebra of the operators. The third describes the size of transition of each vertex, such as the one shown in figure 7.4.

history of a spin network, hence a granular spacetime where the nodes of the graph combine and separate.

This microscopic swarming of quanta, which generates space and time, underlies the calm appearance of the macroscopic reality surrounding us. Every cubic centimetre of space, and every second that passes, is the result of this dancing foam of extremely small quanta.

What is the world made of?

The backdrop of space has disappeared, time has disappeared, classic particles have disappeared, along with the classic fields. So what is the world made of?

The answer now is simple: the particles are quanta of quantum fields; light is formed by quanta of a field; space is nothing more than a field, which is also made of quanta; and time emerges from the processes of this same field. In other words, the world is made entirely from quantum fields (figure 7.8).

These fields do not live *in* spacetime; they live, so to speak, one on top of the other: fields on fields. The space and time that we perceive in large scale are our blurred and approximate image of one of these quantum fields: the gravitational field.

Fields that live on themselves, without the need of a spacetime to serve as a substratum, as a support, and which are capable by themselves of generating spacetime, are called 'covariant quantum fields'. The substance of which the world is made has been radically simplified in recent years. The world, particles, light, energy, space and time – all of this is nothing but the manifestation of a single type of entity: covariant quantum fields.

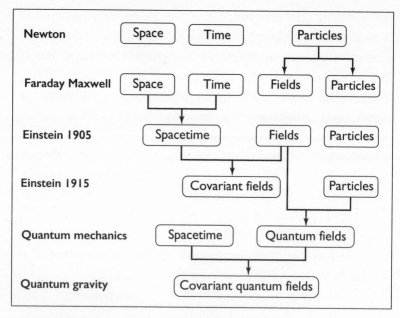

Figure 7.8 What is the world made of? Of only one ingredient: co-variant quantum fields.

Covariant quantum fields have become today the best description that we have of the ἄπειρον, the *apeiron*, the primal substance of which everything is formed hypothesized by the man that could perhaps be called the first scientist and the first philosopher, Anaximander.*

The separation between the curved and continuous space

* '[. . .] all the different elementary particles could be reduced to one universal substance which could equally be called energy or matter, and none of the particles should be privileged and considered more fundamental. This point of view corresponds to Anaximander's doctrine, and I am convinced that in modern physics this is the correct point of view.' Werner Heisenberg, *Physics and Philosophy: The Revolution in Modern Science* (New York, Harper & Row, 1962).

of Einstein's general relativity and the discrete quanta of quantum mechanics which dwell in a flat and uniform space has dissolved. The apparent contradiction is no longer there. Between the spacetime continuum and quanta of space, there is the same relationship as between electromagnetic waves and photons. The waves give an approximate large-scale vision of photons. Photons are the way in which waves interact. Continuous space and time are an approximate large-scale vision of the dynamic of quanta of gravity. The quanta of gravity are the way in which space and time interact. The same mathematics coherently describes the quantum gravitational field as other quantum fields.

The conceptual price paid is the relinquishing of the idea of space, and of time, as general structures within which to frame the world. Space and time are approximations which emerge at a large scale. Kant was perhaps right when he affirmed that the subject of knowledge and its object are inseparable, but he was definitely mistaken when he considered Newtonian space and time as a priori forms of knowledge, parts of an indispensable grammar for understanding the world. This grammar has evolved, and is still in the process of evolving, with the growth of our knowledge.

General relativity and quantum mechanics are, in the end, not as incompatible as they seemed. On closer inspection, they shake hands and engage in a beautiful dialogue. The spatial relations that weave Einstein's curved space are the very interactions weaving the relations between the systems of quantum mechanics. The two become compatible and conjoined, two sides of the same coin, as soon as it is recognized that space and time are aspects of a quantum field, and quantum fields can exist even without being grounded in an external space.

This rarefied picture of the fundamental structure of the physical world is the vision of reality offered today by quantum gravity.

The main reward of this kind of physics is that, as we shall see in the next chapter, infinity disappears. The infinitely small no longer exists. The infinities which plague conventional quantum field theory, predicated on the notion of a continuous space, now vanish, because they were generated precisely by the assumption, physically incorrect, of the continuity of space. The singularities which render Einstein's equations absurd when the gravitational field becomes too strong also disappear: they are only the result of neglecting the quantization of the field. Little by little, the pieces of the puzzle find their place. In the final sections of this book, I describe some of the physical consequences of this theory.

It may appear strange and difficult to think of discrete elementary entities not in space and time, but weaving space and time with their relations. But how strange it must have seemed to listen to Anaximander, when he claimed that beneath our feet there was only the same sky that we can see above our heads? Or to Aristarchus, when he tried to measure the distance from the Earth of the Moon and the Sun, discovering that they are extremely distant, and are therefore not the size of little balls, but gigantic – and the Sun is immense compared to the Earth. Or to Hubble, when he realized that the small, diaphanous clouds between stars are vast seas of immensely distant stars . . .

For centuries, the world has continued to change and expand around us. We see further, understand it better and are astonished by its variety, by the limitations of the images we had of it. The description we manage to produce to account for it becomes increasingly rarefied, yet simple.

We are akin to small, blind moles underground who know little or nothing about the world. But we continue to learn . . .

> But all the story of the night told over
> And all their minds transfigured so together
> More witnesseth than fancy's images
> And grows to something of great constancy;
> But, howsoever, strange and admirable.[2]

PART FOUR
Beyond Space and Time

I have illustrated the basis of quantum gravity, and the image of the world which emerges from it. In the final chapters I describe some consequences of the theory: what the theory tells us about phenomena such as the Big Bang and black holes. I also discuss the current state of possible experiments to test the theory, and what it seems to me nature is telling us – in particular with the failure of the expected observation of supersymmetric particles.

I conclude with a few reflections on what is still missing from our understanding of the world: especially thermodynamics, the role of information in a theory without time and space such as quantum gravity, and the re-emergence of time.

All of this takes us to the edge of what we know, to the vantage point from which we look upon what we definitely don't know, the immense mystery that surrounds us.

8. Beyond the Big Bang

The master

In 1927 a young Belgian scientist, a Jesuit-educated Catholic priest, studies Einstein's equations and realizes – just as Einstein had – that they predict the universe must expand or contract. But instead of foolishly rejecting the result and stubbornly trying to avoid it, as Einstein did, the Belgian priest believes it and looks for astronomical data to test it.

At the time, galaxies were not called galaxies. They were called nebulae because, seen through a telescope, they looked like small, opalescent clouds among the stars. It was not yet known that they are distant, immense islands of stars like our very own galaxy. But the young Belgian priest understands that the scarce available data on the galaxies were indeed compatible with the possibility that the universe is expanding: nearby galaxies are moving away at great speed, as if they had been launched into the sky; distant galaxies are moving away at even greater speed. The universe is swelling like a balloon.

Two years later, the insight is confirmed, thanks to two American astronomers, Henrietta Leavitt and Edwin Hubble. Leavitt discovers a good technique for measuring the distance of the nebulae, confirming that they are very far away, outside of our own galaxy. Using this technique and the great telescope of the Palomar Observatory, Hubble

Figure 8.1 Henrietta Leavitt.

collects precise data that confirm that the galaxies are moving away, at a speed proportional to their distance.

But it is the young Belgian priest who understands, already in 1927, the crucial consequence: if we see a stone flying up, it means that the stone was previously lower down and something has thrown it upwards. If we see the galaxies moving away and the universe expanding, it means that the galaxies were previously much closer and the universe was smaller: and something caused it to start expanding. The young Belgian priest suggests that the universe was originally extremely small and compressed, and started its expansion in a gigantic explosion. He calls this initial state the primordial atom. Today it is known as the Big Bang.

His name was Georges Lemaître. In French, this name sounds like *le maître* meaning 'the master', and few names are more appropriate for the man who first understood the existence of the Big Bang. But in spite of this name, Lemaître's character was reserved; he avoided polemics, and never even claimed priority for the discovery of the expansion of the universe, which ended up being attributed to Hubble. Two episodes from his life illustrate his profound intelligence. The first involves Einstein, the second a pope.

As mentioned, Einstein was sceptical about the expansion of the universe. He had grown up thinking that the universe is fixed, and had not been able to accept the idea that this was not the case. Even the greatest make mistakes and are prey to preconceived ideas. Lemaître met Einstein and tried to dissuade him from his prejudicial view. Einstein resisted, going so far as to answer Lemaître: 'Correct calculations, abominable physics.' Later, Einstein was obliged to recognize that Lemaître was the one who was actually right. It doesn't fall to everyone to disprove Einstein.

Figure 8.2 Georges Lemaître. Copyright Archives Georges Lemaître, Louvain.

The same thing happened again. Einstein had introduced the cosmological constant, the small but important modification of his equations I described in Chapter 3, in the (mistaken) hope of rendering the equations compatible with a static universe. When he had to acknowledge that the universe is not static, he turned against the cosmological constant. Lemaître, for the second time, tried to persuade him to change his mind: the cosmological constant does not render the universe static, but it is nevertheless right, and there is no reason to take it out. On this occasion, too, Lemaître was correct: the cosmological constant produces an acceleration of the expansion of the universe, and this acceleration has recently been measured. Once again, Einstein was wrong and Lemaître was right.

When the idea that the universe had emerged from a Big Bang began to be accepted, Pope Pius XII declared in a public address (on 22 November 1951) that the theory confirmed the account of Creation given in Genesis.[1] Lemaître reacted to this papal position with great concern. He got in touch with the scientific advisor to the pontiff and went to great lengths to persuade the Pope to refrain from making references to links between divine creation and the Big Bang. Lemaître was convinced that it was foolish to mix science and religion in this way: the Bible knows nothing about physics, and physics knows nothing about God.[2] Pius XII allowed himself to be persuaded, and the Catholic church never again made public allusion to the subject. It is not given to everyone to disprove the Pope.

And of course, on this also, it was Lemaître who was right: today there is a great deal of talk concerning the possibility that the Big Bang is not a real beginning, that there could have been another universe before it. Imagine in what an embarrassing position the Catholic Church would find itself today, if Lemaître had not prevented the Pope from making it official doctrine that that Big Bang and Creation were the same thing. *Fiat lux* would have to be changed to 'Switch the light back on!'

To contest both Einstein and the Pope, convincing both that they were mistaken, and to be right in both cases, is surely something of a result. 'The master' lived up to his name.

Today confirmations are overwhelming: the universe, in a far-distant past, was extremely hot and extremely compact, and has expanded since. We can reconstruct in detail the history of the universe, starting with its initial hot, compressed state. We know how atoms, elements, galaxies and stars formed and how the universe as we see it today developed. Recent extended observations of the radiation that fills the universe carried out mainly by the Planck satellite once again confirmed in full the theory of the Big Bang. We know with a reasonable degree of certainty what happened on a large scale to our universe in the last 14 billion years, from the time when it was a ball of fire.

And to think that, initially, the phrase 'theory of the Big Bang' was coined by opponents of the theory, to mock an idea that seemed outlandish . . . Instead, in the end, we were all persuaded: 14 billion years ago the universe was a compressed ball of fire.

But what happened *before* this initial hot and compressed state?

Regressing in time, temperature increases, as does the density of matter and energy. There is a point at which they reach the Planck scale: 14 billion years ago. At that point, the equations of general relativity are no longer valid, because it is no longer possible to ignore quantum mechanics. We enter into the realm of quantum gravity.

Quantum cosmology

To understand what happened 14 billion years ago, therefore, quantum gravity is required. What do the loops tell us about the subject?

Consider an analogous but simpler situation. According to classical mechanics, an electron falling straight into an atomic nucleus would be swallowed by the nucleus and disappear. But this is not what happens in reality. Classic mechanics is incomplete, and it is necessary to take quantum effects into account. A real electron is a quantum object and does not follow a precise trajectory: it isn't possible to keep it inside too small a region. The more it is concentrated, the more it slips rapidly away. If we want to stop it around the nucleus, the most we can do is to force it into an orbit of the size of the smallest atomic orbital: it could not stay any closer to the nucleus. Quantum mechanics prevents a real electron from falling into a nucleus. A quantum repulsion pushes away the electron when it gets too close to the centre. Thus, thanks to quantum mechanics, matter is stable. Without it, electrons would fall into nuclei, there would be no atoms and we would not exist.

The same applies to the universe. Let us imagine a universe contracting and becoming extremely small, squashed by its own weight. According to Einstein's equations, this universe would be squashed ad infinitum and at a certain point would disappear altogether, like the electron falling into the nucleus. This is the Big Bang predicted by Einstein's equations, if we ignore quantum theory.

But if we take quantum mechanics into account, the universe cannot be indefinitely squashed. A quantum repulsion makes it rebound. A contracting universe does not collapse down to a point: it bounces back and begins to expand, as if it were emerging from a cosmic explosion (figure 8.3).

Figure 8.3 The Big Bounce of the universe in a graphical representation by Francesca Vidotto, the Italian scientist who first used spinfoams to compute the probability of this process.

The past of our universe may therefore well be the result of just such a rebound. A gigantic rebound known as a Big Bounce instead of Big Bang. This is what seems to emerge from the equations of loop quantum gravity when they are applied to the expansion of the universe.

The image of the bounce must not be taken literally. Going back to the example of the electron, recall that if we want to place an electron as close as possible to an atom, the electron is no longer a particle; we can think of it, instead, as opened up in a cloud of probabilities. An exact position no longer makes sense for the electron. The same for the universe: in the crucial passage through the Big Bounce, we can no longer think of a single, although granular, space and time, but only of a spread-out cloud of probabilities in which time and space wildly fluctuate. At the Big Bounce, the world is dissolved into a swarming cloud of probabilities, which the equations still manage to describe.

Our universe could thus be the result of the collapse of a previous contracting universe passing across a quantum phase, where space and time are dissolved into probabilities.

The word 'universe' becomes ambiguous. If, by 'universe', we mean 'all that there is', then, by definition, there cannot be a second universe. But the word 'universe' has assumed another meaning in cosmology: it refers to the spacetime continuum that we see directly around us, filled with galaxies the geometry and history of which we observe. There is no reason to be certain that, in *this* sense, this universe is the only one in existence. We can reconstruct the past up to the time when, as in the image by John Wheeler, the spatiotemporal continuum breaks up like sea foam and fragments into a quantum cloud of probabilities, and there is no reason to discard the possibility that beyond this hot foam there could not be another spatiotemporal continuum, similar to the one which we perceive around us.

The probability for a universe to cross the phase of the Big Bounce, passing from contraction to expansion, can be computed using the techniques described in the preceding chapter: the spacetime boxes. Calculations are made using spinfoams that connect the contracting universe with the expanding one.

All of this is still at an exploratory stage, but what is remarkable in this story is that today we have equations with which to try to describe these events. We are beginning to cast the first few cautious glances, for the moment only theoretically, beyond the Big Bang.

9. Empirical Confirmations?

The appeal of quantum cosmology goes beyond the fascinating theoretical explorations of what there might be beyond the Big Bang. There is another reason for studying the application of the theory to cosmology: it might provide the opportunity to find out whether or not the theory is actually correct.

Science works because, after hypotheses and reasoning, after intuitions and visions, after equations and calculations, we can check whether we have done well or not: the theory gives predictions about things we have not yet observed, and we can check whether these are correct, or not. This is the power of science, that which grounds its reliability and allows us to trust in it with confidence: we can check whether a theory is right or wrong. This is what distinguishes science from other kinds of thinking, where deciding who is right and who is wrong is usually a much thornier question, sometimes even devoid of meaning.

When Lemaître defends the idea that the universe is expanding, and Einstein does not believe it, one of the two is wrong; the other right. All of Einstein's results, his fame, his influence on the scientific world, his immense authority, count for nothing. The observations prove him wrong, and it's game over. An obscure Belgian priest is right. It is for this reason that scientific thinking has power.

The sociology of science has shed light on the complexity of the process of scientific understanding; like any other human endeavour, this process is beset by irrationality, intersects with the game of power and is affected by every sort of social and cultural influence. Nevertheless, despite all of this, and in opposition to the exaggerations of a few postmodernists, cultural relativists and the like, none of this diminishes the practical and theoretical efficacy of scientific thinking. Because in the end, in the majority of cases, it is possible to establish with clarity who is right and who is wrong. And even the great Einstein could go on to say (and he did so), 'Ah . . . I made a mistake!' Science is the best strategy if we value reliability.

This does not mean that science is just the art of making measurable predictions. Some philosophers of science overly circumscribe science by limiting it to its numerical predictions. They miss the point, because they confuse the instruments with the objectives. Verifiable quantitative predictions are instruments to validate hypotheses. The objective of scientific research is not just to arrive at predictions: it is to understand how the world functions; to construct and develop an image of the world, a conceptual structure to enable us to think about it. Before being technical, science is visionary.

The verifiable predictions are the sharpened tool which allows us to find out when we have misunderstood something. A theory lacking empirical confirmation is a theory which has not yet passed its exams. Exams never end, and a theory is not completely confirmed by one, two or three experiments. But it progressively acquires credibility, stage by stage, as its predictions are revealed to be correct. Theories such as general relativity and quantum mechanics, which initially left many perplexed, earned their credibility gradually, as all of their predictions – even the most bizarre – were gradually confirmed by experiments and observations.

The importance of experimental proof, on the other hand, does not mean that, without new experimental data, we cannot make advances. It is often said that science takes steps forward only when there is new experimental data. If this were true, we would have little hope of finding the theory of quantum gravity before measuring something new, but this is patently not the case. Which new data were available to Copernicus? None. He had the same data as Ptolemy. Which new data did Newton have? Almost none. His real ingredients were Kepler's laws and Galileo's results. What new data did Einstein have to discover general relativity? None. His ingredients were special relativity and Newton's theory. It simply isn't true that physics advances only when it is afforded new data.

What Copernicus, Newton, Einstein and many others did was to build upon pre-existing theories which synthesized empirical knowledge across vast fields of nature, and to find a way of combining and rethinking them to improve the general picture.

This is the basis on which the best research on quantum gravity operates. The origin of knowledge, as always in science, is ultimately empirical. But the data on which quantum gravity is built is not new experiments: it is the theoretical edifices which have already structured our knowledge of the world, in forms which are only partly coherent. The 'experimental data' for quantum gravity are general relativity and quantum mechanics. Building on these, trying to understand how a world in which both quanta and curved space exist may be made coherent, we attempt to look towards the unknown.

The enormous success of the giants who have preceded us in similar situations, such as Newton, Einstein and Dirac, gives us encouragement. We do not presume to be of their stature. But we have the advantage of sitting on their shoulders, and this allows us to look further than they did. One way or another, we cannot but try.

We must distinguish between clues and strong evidence. Clues are what set Sherlock Holmes on the right track, allowing him to solve a mysterious case. Strong evidence is what the judge needs to sentence the guilty. Clues put us on the right path towards a correct theory. Strong evidence is that which subsequently allows us to trust whether the theory we have built is a good one or not. Without clues, we search in the wrong directions. Without evidence, a theory is not reliable.

The same applies to quantum gravity. The theory is in its infancy. Its theoretical apparatus is gaining solidity, and the fundamental ideas are being clarified: the clues are good, and concrete – confirmed predictions are still missing. The theory has not yet taken its exams.

Signals from nature

The most studied alternative to the research direction recounted in this book is string theory. The majority of physicists who have worked on string theory, or string-related theories, expected that as soon as the new particle accelerator at CERN in Geneva began to function (the LHC or Large Hadron Collider), particles of a new kind never before observed, but anticipated by the theory, would immediately become evident: supersymmetric particles. String theory needs these particles to be consistent: that is why the string theorists eagerly expected them to be found. Loop quantum gravity, on the other hand, is well defined even without supersymmetric particles. The loop theorists were inclined to think that these particles might not exist.

The supersymmetric particles were not observed, to the great disappointment of many. The fanfare that greeted the discovery of the Higgs bosun in 2013 also masked this disappointment. The supersymmetric particles are not there at the energy where many string theorists expected them to be. This is not a definitive proof of anything – far from it; but nature has given a small clue in favour of the loops.

There have been three major experimental results in fundamental physics in recent years. The first is the revelation of the Higgs boson at CERN in Geneva (figure 9.1). The second is the measurements made by the Planck satellite (figure 9.2), measurements, the data of which were also made public in 2013, confirming the standard cosmological model. The third is the first detection of gravitational waves announced

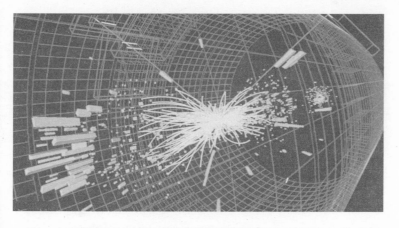

Figure 9.1 An event at CERN which shows the formation of the Higgs particle.

Figure 9.2 The Planck satellite.

in the first months of 2016. These are the three signals that nature has recently given us.

There is something in common between these three results: the complete absence of surprise. This does not diminish their importance: if anything, it makes them even more meaningful. The discovery of the Higgs boson is a rock-hard confirmation of the validity of the ideas behind the standard model of elementary particles, based on quantum mechanics. It is the verification of a prediction made thirty years previously. The Planck measurements are a solid confirmation of the standard cosmological model, based on general relativity with the cosmological constant. The detection of gravitational waves is a spectacular confirmation of general relativity, a theory a hundred years old. The three results, obtained with strenuous technological efforts and extensive collaborations between hundreds of scientists, do nothing other than reinforce the understanding that we already had of the structure of the universe. No real surprises.

But such an absence of surprises was in a sense itself surprising, because many expected to be surprised, that is, to see 'new physics', not yet described by established theories. They expected supersymmetry at CERN, not the Higgs boson. And many expected that Planck would measure discrepancies from the standard cosmological model, discrepancies that would support alternative cosmological theories to general relativity.

But no. What nature is confirming is simple: general relativity, quantum mechanics and, within quantum mechanics, the standard model.

Many theoretical physicists are today looking for new theories by picking arbitrary hypotheses. 'Let us imagine that . . .' I don't think that this way of doing science has ever produced good results. Our fantasy is too limited to 'imagine' how the world may be made, unless we search for inspiration in the traces we have at our disposal. The traces that we have – our clues – are either the theories which have been successful, or new experimental data, nothing else. It is in this data and in these theories that we must try to uncover what we have been unable yet to imagine. This is how Copernicus, Newton, Maxwell and Einstein proceeded. They never tried to 'guess' a new theory – unlike, in my opinion, the way in which too many theoretical physicists are trying to do today.

The three recent experimental results I mentioned speak with the voice of Nature itself: 'Stop dreaming of new fields and strange particles; supplementary dimensions, other symmetries, parallel universes, strings, and whatever else. The pieces of the puzzle are simpler: general relativity, quantum mechanics and the standard model. The next step forward may be "only" a question of combining them in the correct manner.' It's reassuring advice for the loop quantum gravity community, because these are the hypotheses of the theory: general relativity, quantum mechanics and compatibility with the standard model, nothing else. The radical conceptual consequences – the quanta of space, the disappearance of time – are not bold hypotheses: they are the rational consequences that follow from taking the basic insights of our best theories seriously.

Once again, these are not definitive proofs. Supersymmetric particles might finally exist, perhaps, at a scale still not reached, and could exist even if loop theory is correct. Supersymmetry failed to show up where expected, and string theorists are a little downcast, loop theorists are buoyant, but it is still a matter of clues; there is no strong evidence at all.

To find more concrete confirmation of the theory, we need to look elsewhere. The primordial universe could open the window to predictions capable of confirming the theory. In a not too distant future, we hope. Or they could prove the theory wrong.

A window on to quantum gravity

If we have the equations that describe the transition of the universe across the quantum phase, we can compute effects of quantum phenomena upon the universe which we observe today. The universe is filled with cosmic radiation: a sea of photons remained in the cosmos since the early hot phase, the residual glare of the early high temperature.

The electromagnetic field in the immense space between galaxies trembles like the surface of the sea after a big storm. This quivering, disseminated throughout the universe, is called the cosmic background radiation. It has been studied in the past few years by satellites such as COBE, WMAP and, most recently, Planck. An image of the minute fluctuations of this radiation is given in figure 9.3. The details of the structure of this radiation tell us the history of the universe and, hidden in the folds of these details, there could be footprints of the quantum beginning of our universe.

One of the most active sectors of research in loop quantum gravity is studying how the quantum dynamic of the primordial universe is reflected in this data. The results are preliminary, but encouraging. With more calculations and more precise measurements, it should be possible to arrive at a test of the theory.

In 2013 Abhay Ashtekar, Ivan Agullo and William Nelson published an article in which they calculate that, under certain hypotheses, the statistical distribution of the fluctuations of this source of cosmic radiation should reveal the effect of

Figure 9.3 The fluctuations of the cosmic background radiation. This is the image of the oldest object in the universe available to us. These fluctuations were produced 14 billion years ago. In the statistics of such fluctuations we hope to find confirmation of the predictions of quantum gravity.

the initial bounce: the wide-angle fluctuations should be different from those predicted by the theory that does not take quanta into account. The current state of the measurement is described in figure 9.4, where the black line represents the prediction by Ashtekar, Agullo and Nelson, and the grey dots the measured data. For now, these are not sufficient to evaluate whether the upward bend of the black line predicted by the three authors is correct or not. But measurements are getting more precise. The situation is still fluid. But those who, like myself, have spent their lives seeking to understand the secrets of quantum space are following with close attention, anxiety and hope the continuous honing of our capacity to make observations, to measure and to calculate – and are awaiting the moment in which nature will tell us whether we are right or not.

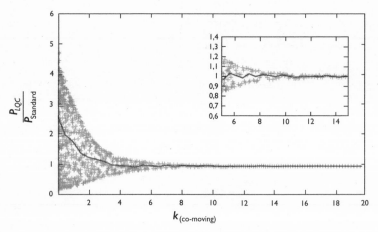

Figure 9.4 Possible prediction of the spectrum of background radiation, of loop quantum gravity (shown by the solid line) compared with the current experimental errors (as represented by the points). Courtesy of A. Ashtekar, I. Agullo and W. Nelson.

Traces of the great primordial heat must also be in the gravitational field itself. The gravitational field, too, that is to say, space itself, must be tremulous like the surface of the sea. Therefore, a cosmic *gravitational* background radiation must also exist – older even than the electromagnetic one, because the gravitational waves are disturbed less by matter than the electromagnetic ones and were able to travel undisturbed even when the universe was too dense to let the electromagnetic waves pass.

We have now observed gravitational waves directly, with the LIGO detector, formed by two arms of a few miles in length, at a right angle to each other, in which laser beams measure the distance between three fixed points. When a gravitational wave passes, the space lengthens and shortens imperceptibly, and the lasers reveal this minuscule variation.* The gravitational waves observed were generated by an astrophysical event: colliding black holes. These are phenomena described by general relativity which do not involve quantum gravity. But a more ambitious experiment called LISA is at the stage of being evaluated and is capable of doing the same thing but on a much larger scale: by putting into orbit three satellites, not around the Earth but around the Sun, as if they were miniature planets tracking the Earth in its orbit. The three satellites are connected by laser beams measuring the distance between them or, better still, the variations in the distances when a gravitational wave passes. If LISA is launched, it should be able to see not only the gravitational waves produced by stars and black holes but also the

---

* This is an interferometer: it uses the interference between the lasers which run along the two arms to reveal the minute variations in length of these arms.

diffuse background of primordial gravitational waves generated at a time close to the Big Bang. These waves should tell us about the quantum bounce.

In the subtle irregularities of space, we should be able to find traces of events which took place 14 billion years ago, at the origin of our universe, and confirm our deductions on the nature of space and time.

10. Quantum Black Holes

Black holes populate our universe in great number. They are regions in which space is so curved as to collapse in on itself, and where time comes to a standstill. As mentioned, they form, for instance, when a star has burned up all of the available hydrogen and collapses.

Frequently, the collapsed star formed part of a pair of neighbouring stars and, in this case, the black hole and the surviving counterpart circle one around the other; the black hole sucks matter from the other star continuously (as in figure 10.1).

Astronomers have found many black holes with a size

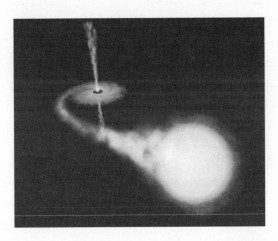

Figure 10.1 Representation of a couple star/black hole.
The star loses matter, which is partly absorbed by the black hole, partly projected by it in jets in the direction of its poles.

(that is, mass) of the order of our Sun (a bit larger, in fact). But there are also gigantic back holes. There is one of these at the centre of almost all of the galaxies, including our own.

The black hole at the centre of our own galaxy is currently being studied in detail. It has a mass a million times greater than our Sun. Every so often, a star gets too close to this monster, is disintegrated by the gravitational distortion and swallowed by the cyclopean black hole, like a small fish swallowed by a whale. Imagine a monster the size of a million Suns, which swallows in an instant our own Sun and its miniature planets . . .

There is a wonderful ongoing project to construct a network of radio antennae distributed across the Earth from pole to pole, with which astronomers will be able to achieve a resolution sufficient to 'see' the galactic black hole. What we expect to see is a small black disc surrounded by the light produced by the radiation of the matter falling in.

What enters a black hole does not come out again, at least if we neglect quantum theory. The surface of a black hole is like the present: it can be crossed only in one direction. From the future, there is no return. For a black hole, the past is the outside; the future is the inside. Seen from outside, a black hole is like a sphere which can be entered but out of which nothing can come. A rocket could stay positioned at a fixed distance from this sphere, which is called the *horizon* of the black hole. To do so it needs to keep its engines firing intensely, to resist the gravitational pull of the hole. The powerful gravity of the hole implies that time slows down for this rocket. If the rocket stays near enough to the horizon for one hour, and then moves away, it would then find that, outside, in the meantime, centuries have passed. The closer the rocket stays to the horizon, the slower – with respect to the outside – time runs for it. Thus, travelling to the past is difficult, but travelling to the future is easy: we need only to get close to a black hole with a spaceship, keep within its vicinity for a while, and then move away.

On the horizon itself, time stops: if we get extremely close to it and then move away after a few of our minutes, a million years might have elapsed in the rest of the universe.

The really surprising thing is that the properties of these strange objects, today commonly observed, were *foreseen* by Einstein's theory. Now, astronomers study these objects in space, but until not long ago black holes were considered a scarcely credible and bizarre consequence of an outlandish theory. I remember my university professor introducing them as solutions to Einstein's equations, to which 'real objects were unlikely ever to correspond'. This is the stupendous capacity of theoretical physics to discover things before they are observed.

The black holes we observe are well described by Einstein's theory, and quantum mechanics is not needed to understand them. But there are two mysteries of black holes that do require quantum mechanics in order to be unravelled and, for each of these, loop theory offers a possible solution. One of these could also offer an opportunity to test the theory.

The first application of quantum gravity to black holes concerns a curious fact discovered by Stephen Hawking. Early in the 1970s he theoretically deduced that black holes are 'hot'. They behave like hot bodies: they emit heat. In doing so, they lose energy and hence mass (since energy and mass are the same thing), becoming progressively smaller. They 'evaporate'. This 'evaporation of black holes' is the most important discovery made by Hawking.

Objects are hot because their microscopic constituents

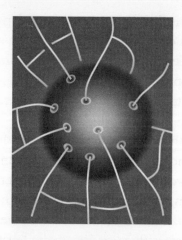

Figure 10.2 The surface of a black hole crossed by loops, that is to say, by links of the spin network that describe the state of the gravitational field. Each loop corresponds to a quantum area of the black hole's surface. © John Baez.

move. A hot piece of iron, for example, is a piece of iron where the atoms vibrate very rapidly around their equilibrium position. Hot air is air in which molecules move faster than in cold air.

What are the elementary 'atoms' that vibrate, making a black hole hot? Hawking left this problem unanswered. Loop theory provides a possible answer. The elementary atoms of a black hole that vibrate, and are thus responsible for its temperature, are the individual quanta of space on its surface.

Thus, it is possible to understand the peculiar heat of black holes predicted by Hawking using loop theory: the heat is the result of the microscopic vibrations of the individual atoms of space. These vibrate because in the world of quantum mechanics *everything* vibrates; nothing stays still. The impossibility of anything being entirely and continuously still in a place is at the heart of quantum mechanics.

Black-hole heat is directly connected to loop quantum gravity's fluctuations of the atoms of space. The precise position of the black hole's horizon is determined only in relation to these microscopic fluctuations of the gravitational field. Hence, in a certain sense, the horizon fluctuates like a hot body.

There is another way of understanding the origin of the heat of black holes. The quantum fluctuations generate a correlation between the interior and the exterior of a hole. (I will speak at length about correlations and temperature in Chapter 12). Quantum uncertainty across the horizon of the black hole generates fluctuations of the horizon's geometry. But fluctuations imply probability, and probability implies thermodynamics, and therefore temperature. Concealing from us a part of the universe, a black hole makes its quantum fluctuations detectable in the form of heat.

It was a young Italian scientist, Eugenio Bianchi, today a professor in the United States, who completed an elegant calculation which shows how, starting from these ideas and from the basic equations of loop quantum gravity, it is possible to derive the formula for the heat of black holes foreseen by Hawking (figure 10.3).

The second application of loop quantum gravity to black-hole physics is more spectacular. Once collapsed, a star vanishes

Figure 10.3 Stephen Hawking and Eugenio Bianchi. On the blackboard are the principal equations of loop quantum gravity which describe black holes.

from external view: it is inside the black hole. But, inside the hole, what happens to it? What would you see if you let yourself fall into the hole?

At first, nothing in particular: you would cross the surface of the black hole without major injuries – then you would plummet towards the centre, at ever greater speed. And then? General relativity predicts that everything is squashed at the centre into an infinitely small point of infinite density. But this is, once again, if we ignore quantum theory.

If we take quantum gravity into account, this prediction is no longer correct – there is quantum repulsion – the same repulsion that makes the universe bounce at the Big Bang. What we expect is that, on getting closer to the centre, the falling matter is slowed down by this quantum pressure, up to a very high but finite density. Matter gets squashed, but not all the way to an infinitely small point, because there is a limit to how small things can be. Quantum gravity generates a huge pressure that makes matter bounce out, precisely as a collapsing universe can bounce out into an expanding universe.

The bounce of a collapsing star can be very fast, if watched from down there. But – remember – time passes much more slowly there than outside. Seen from the outside, the process of the bounce can take billions of years. After this time, we can see the black hole explode. In the end, basically, this is what a black hole is: a shortcut to the distant future.

Thus, quantum gravity might imply that black holes are not eternally stable objects, as classical general relativity predicted, after all. They are, ultimately, unstable.

Seeing these black-hole explosions would be a spectacular confirmation for the theory. Very old black holes, such as those formed in the early universe, could be exploding today.

Some recent calculations suggest that the signals of their explosion could be in the range of radio telescopes. It has even been suggested that certain mysterious radio pulses which radio astronomers have *already* measured, called Fast Radio Bursts, could be, precisely, signals generated by the explosion of primordial black holes. If this was confirmed, it would be fantastic: we would have a direct sign of a quantum gravitational phenomenon. Let's wait and see . . .

11. The End of Infinity

When we take quantum gravity into account, the infinite compression of the universe into a single, infinitely small point predicted by general relativity at the Big Bang disappears. Quantum gravity is the discovery that no infinitely small point exists. There is a lower limit to the divisibility of space. The universe cannot be smaller than the Planck scale, because nothing exists which is smaller than the Planck scale.

If we ignore quantum mechanics, we ignore the existence of this lower limit. The pathological situations predicted by general relativity, where the theory gives infinite quantities, are called singularities. Quantum gravity places a limit to infinity, and 'cures' the pathological singularities of general relativity.

The same happens at the centre of black holes: the singularity that classic general relativity anticipated disappears as soon as we take quantum gravity into account.

There is another case, of a different kind, in which quantum gravity places a limit to the infinite, and it regards forces such as electromagnetism. Quantum field theory, started by Dirac and completed in the 1950s by Feynman and his colleagues, describes these forces well but is full of mathematical absurdities. When we use it to compute physical processes, we often obtain results which are infinite, and mean nothing. They are called divergences. The divergences are then eliminated with calculations, using a baroque technical procedure which leads to finite final results. In practice, it works, and the numbers, in the end, come out right; they reproduce the experimental measurements. But why must the theory go via the infinite to arrive at reasonable numbers?

In the last years of his life, Dirac was very unhappy with the infinities in his theory and felt that, all things considered, his objective of truly understanding how things worked was not achieved. Dirac loved conceptual clarity, even if what was clarity to him was not always clarity to others. But infinities do not make for clarity.

But the infinities of quantum field theory follow from an assumption at the basis of theory: the infinite divisibility of space. For example, to calculate the probabilities of a process, we sum up – as Feynman has taught us – all of the ways in which the process could unfold, and these are infinite, because they can happen in any one of the infinite points of a spatial continuum. This is why the result can be infinite.

When quantum gravity is taken into account, these infinities also disappear. The reason is clear: space is not infinitely divisible, there are no infinite points; there are no infinite things to add up. The granular discrete structure of space resolves the difficulties of the quantum theory of fields, eliminating the infinities by which it is afflicted.

This is a tremendous result: on the one hand, taking quantum mechanics into account resolves the problems generated by the infinities of Einstein's theory of gravity, that is to say, the singularities. On the other, taking gravity into account solves the problems generated by quantum field theory, that is to say, the divergences. Far from being contradictory, as they at first seemed, the two theories each offer the solution to the problems posed by the other!

Putting a limit to infinity is a recurrent theme in modern

physics. Special relativity may be summarized as the discovery that there exists a maximum velocity for all physical systems. Quantum mechanics can be summarized as the discovery that there exists a maximum of information for each physical system. The minimum length is the Planck length L_P, the maximum velocity is the speed of light c, and the total information is determined by the Planck constant h. This is summarized in table 11.1.

The existence of these minimum and maximum values for length, velocity and action fixes a natural system of units. Instead of measuring speed in kilometres per hour, or in metres per second, we can measure it in fractions of the speed of light. We can fix the value 1 for the velocity c and write, for example, $v = \frac{1}{2}$, for a body which is moving at half the speed of light. In the same way, we can posit by definition and measure length in multiples of Planck's length. And we can posit $h = 1$ and measure actions in multiples of Planck's constant. In this way, we have a natural system of fundamental unities from which the others follow. The unity of time is the time that light takes to cover the Planck length, and so on. The natural unities are commonly used in research on quantum gravity.

The identification of these three fundamental constants

Table 11.1 Fundamental limitations discovered by theoretical physics.

Physical Quantity	Fundamental constant	Theory	Discovery
Velocity	c	Special relativity	A maximum velocity exists
Information (actions)	h	Quantum mechanics	A minimum of information exists
Length	L_P	Quantum gravity	A minimum length exists

places a limit to what seemed to be the infinite possibilities of nature. It suggests that what we call infinite often is nothing more than something which we have not yet counted, or understood. I think this is true in general. 'Infinite', ultimately, is the name that we give to what we do not yet know. Nature appears to be telling us that there is nothing truly infinite.

There is another infinity which disorientates our thinking: the infinite spatial extension of the cosmos. But as I illustrated in Chapter 3, Einstein has found the way of thinking of a finite cosmos without borders. Current measurements indicate that the size of the cosmos must be larger than 100 billion light years. This is the order of magnitude of the universe we have indirect access to. It is around 10^{120} times greater than the Planck length, a number of times which is given by a 1 followed by 120 zeroes. Between the Planck scale and the cosmological one, then, there is the mind-blowing separation of 120 orders of magnitude. Huge. Extraordinarily huge. But *finite*.

In this space – between the size of the minute quanta of space, up to quarks, protons, atoms, chemical structures, mountains, stars, galaxies (each formed by one hundred billion stars), clusters of galaxies, and right up until the seemingly boundless visible universe of more than 100 billion galaxies – unfolds the swarming complexity of our universe; a universe we know only in a few aspects. Immense. Finite.

The cosmological scale is reflected in the value of the cosmological constant Λ, which enters into the basic equations of our theories. The fundamental theory contains, therefore, a very large number: the ratio between the cosmological constant and the Planck length. It is this large number that opens the way to the vast complexity of the world. But what we see and understand of the universe is not an infinity to drown in. It is a wide sea, but a finite one.

The book of *Ecclesiasticus*, or Sirach,* opens with a stupendous question:

> Who can number the sand of the sea, and the drops of rain, and the days of eternity? Who can find out the height of heaven, and the breadth of the earth, and the deep, and wisdom?

Not much longer after these lines were composed, another great text was written, with an opening which still resounds:

> Some think, O King Hiero, that the grains of sand cannot be counted.

This is the opening of *Psammites* (*The Sand Reckoner*) by Archimedes, in which the greatest scientist of antiquity . . . counts the grains of sand in the universe!

He does so in order to demonstrate that their number is large but finite, and can be determined. The numerical system of antiquity did not allow for dealing with very large numbers easily. In *The Sand Reckoner*, Archimedes develops a new system of numbering, similar to our exponentials, that makes it possible to deal with very large numbers, and shows its power by counting (certainly playfully) how many grains

---

* The *Ecclesiasticus* is considered part of the Bible by Catholics, most of the Oriental Orthodox Church and some Jews. The Lutheran churches include it in their lectionaries, and as a book proper for reading, devotion and prayer, but not in the Bible. For most Jews and the Anglican Church the situation is similar.

of sand there are, not just on the seashores but in the entire universe.

The Sand Reckoner is playful, but profound. With a flight of fancy that seems to anticipate the Enlightenment by millennia, Archimedes rebels against the form of knowledge that insists on there being mysteries which are *intrinsically* inaccessible to human thought. He does not claim to know the exact dimensions of the universe, or the precise number of grains of sand. It isn't the completeness of his knowledge that he is asserting. On the contrary, he is explicit about the approximate and provisional nature of his estimates. He speaks about possible alternatives regarding the true size of the universe – between which he does not make a definite choice. The point at stake here is not the presumption of knowing everything. It is the opposite: an awareness that yesterday's ignorance may have light shed on it today, and that today's might be illuminated tomorrow.

The central point is rebellion against the renunciation of the desire to know: a declaration of faith in the comprehensibility of the world, a proud retaliation to those who remain satisfied with their own ignorance, who call infinite that which we don't understand and delegate knowledge to elsewhere.

Centuries have passed, and the text of *Ecclesiasticus*, along with the rest of the Bible can be found in countless homes, while Archimedes' text is read only by the few. Archimedes was slaughtered by the Romans during the sacking of Syracuse, the last proud remnant of Magna Grecia to fall under the Roman yoke, during the expansion of that future empire which would soon adopt *Ecclesiasticus* as one of the foundational texts of its official religion, a position which it was to occupy there for more than a thousand years. During that millennium, the calculations made by Archimedes languished in a state of incomprehensibility: no one was able to use, or even to understand them.

Near Archimedes' Syracuse there is one of the most beautiful sites in Italy, the theatre of Taormina, which looks out at the Mediterranean and upon Mount Etna, the smoking volcano. In Archimedes' time, the theatre was used to stage plays by Sophocles and Euripides. The Romans adapted it for gladiatorial combat, for the pleasure of watching gladiators die.

The sophisticated playfulness of *The Sand Reckoner* is perhaps not only about an audacious mathematical construction, or the virtuosity of one of the most extraordinary minds of antiquity. It is also a defiant cry of reason, which recognizes its own ignorance but refuses to delegate to others the source of knowledge. It is a small, reserved and powerfully intelligent manifesto against infinity – against obscurantism.

Quantum gravity is one of the many lines that continue the quest of *The Sand Reckoner*. We are counting the grains of space of which the cosmos is made. A vast cosmos, but a finite one.

The only truly infinite thing is our ignorance.

12. Information

We are approaching the conclusion of our journey. In the previous few chapters, I spoke about the concrete applications of quantum gravity: the description of what happened to the universe around the time of the Big Bang; the description of the properties of the heat of black holes and the suppression of infinity.

Before concluding, I would like to return to the theory, but looking at its future, and to speak about information: a spectre that is haunting theoretical physics, arousing enthusiasm and confusion.

This chapter is different from the preceding ones, where I spoke of ideas and theories not yet tested but well defined; here, I'm speaking of ideas still confused, badly in need of organization. If, dear reader, you have found the journey so far a little rough, then hold on tighter, because we're now flying between voids of air. If this chapter seems particularly opaque, it's not because your ideas are confused. It's because the one with the confused ideas is me.

Many scientists suspect today that the concept of 'information' may turn out to be a key for new advances in physics. Information is mentioned in the foundations of thermodynamics, the science of heat, the foundation of quantum mechanics and in other areas besides, with the word quite often used very imprecisely. I believe there is something important in this idea. I'll try to explain why, and to show what information has to do with quantum gravity.

Before anything else, what *is* information? The word 'information' is used in common parlance to mean a variety of different things, and this imprecision is a source of confusion in science as well. The scientific notion of information, however, was defined with clarity in 1948, by the American mathematician and engineer Claude Shannon, and is something very simple: information is the measure of the number of possible alternatives for something. For example, if I throw a die, it can land on one of six faces. When we've seen it fall on a particular one of these, we have an amount of information $N = 6$, because the possible alternatives are six in number. If I don't know which day of the year is your birthday; there are 365 distinct possibilities. If you tell me the date, I have the information $N = 365$. And so on.

Instead of the number of alternatives N, scientists measure information in terms of a quantity called S, for 'Shannon information'. S is defined as the logarithm in base 2 of N: $S = \log_2 N$. The advantage of using the logarithm is that the unit of measurement $S = 1$ corresponds to $N = 2$ (because $1 = \log_2 2$), making the unit of information the minimum number of alternatives: the choice between *two* possibilities. This unit of measurement is called 'bit'. When I know at roulette that a red number has come up rather than a black, I have one bit of information; when I know that a *red, even* number has won, I have two bits of information; when an even red number 'manque' (eighteen or less, in roulette parlance) wins, I have three bits. Two bits of information correspond to four alternatives (red even, red uneven, black even, black uneven). Three bits of information correspond to eight alternatives. And so on.*

* A subtle point: information doesn't measure what I know but the number of possible alternatives. The information I am given when the

A key point is that information can be located somewhere. Imagine, for instance, that you have in your hand a ball which can be either black or white. Imagine that I also have a ball which can be either black or white. There are two possibilities on my part, and two on yours. The total number of possibilities is four (2 x 2): white-white; white-black; black-white and black-black. Now, suppose that for some reason we are certain that the two balls are opposite in colour (for instance, because we have taken the balls from a box that contained only one white and one black ball). The total number of alternatives is then *only 2* (white-black or black-white), even if the alternatives are still two on my part and two on yours. Note that, in this situation, something peculiar happens: if you look at your ball, then *you know the colour of mine*. In this case, we say that the colours of the two balls are correlated, that is to say, linked to one another. We say that my ball 'has information' about yours (as well as vice versa).

If you think about it, this is precisely what happens in life when we communicate: for example, when I phone you, I know that the phone causes the sounds on your side to be dependent on the sounds on mine. The sounds on both sides are linked, like the colour of the balls.

The example is not chosen at random: Shannon, who invented the theory of information, worked for a telephone company, and was looking for a way to measure accurately how much a telephone line could 'carry'. But *what* does a

---

number 3 comes up in roulette is $N = 37$, because there are 37 numbers; but the information I have when number 3 wins on red is $N = 18$, because there are 18 red numbers. How much information do we have if we learn which of the brothers Karamazov murdered their father? The answer depends on how many Karamazov brothers there are.

telephone line carry? It carries information. It carries the capacity to distinguish between alternatives. For this reason, Shannon defined information.

Why is the notion of information useful, perhaps even fundamental, to understanding the world? For a subtle reason: because it measures the ability of one physical system to communicate with another physical system.

Let's return for a final time to the atoms of Democritus. Let's imagine a world formed of an interminable sea of atoms which bounce, attract and cling together, and of nothing else. Aren't we missing something?

Plato and Aristotle insisted on the fact that something was indeed missing; they thought that the *form* of things was this something *extra* that had to be added to the *substance* of which things were made in order to understand the world. For Plato, forms exist by themselves, in an ethereal ideal world of forms, a world of 'ideas'. The idea of a horse exists prior to and independently of any actual horse. For Plato, a real horse is nothing but a pale reflection of the idea of a horse. The atoms which make up the horse count for little: what counts is the 'horseness', the abstract form. Aristotle is a bit more realistic, but for him, too, the form cannot be reduced to the substance. In a statue, there is more than the stone of which it is made. This *more*, for Aristotle, is the form. This is the basis of the critique of Democritus's materialism in antiquity. It still remains a common critique of materialism.

But was Democritus really proposing that everything can be reduced to atoms? Let's look at it more closely. Democritus says that when atoms combine what counts is their form, their arrangement in the structure, as well as the way in which they *combine*. He gives the example of the letters of the alphabet: there are only twenty or so letters but, as he puts it, 'It is possible for them to combine in diverse modes, in order to produce comedies or tragedies, ridiculous stories or epic poems.'

There are more than just atoms in this idea: what counts is the *way* in which they are combined, one in relation to another. But what relevance can the way in which they are combined have, in a world in which there is nothing but other atoms?

If the atoms are also an alphabet, who is able to read the phrases written with this alphabet?

The answer is subtle: the way in which the atoms arrange themselves is correlated with the way *other* atoms arrange themselves. Therefore, a set of atoms can have *information*, in the technical, precise sense described above, about another set of atoms.

This, in the physical world, happens continuously and throughout, in every moment and in every place: the light which arrives at our eyes carries information about the objects which it has played across; the colour of the sea has information on the colour of the sky above it; a cell has information about the virus that is attacking it; a new living being has plenty of information because it is correlated with its parents, and with its species; and you, dear reader, when reading these lines, receive information about what I am thinking while writing them, that is to say, about what is happening in my mind at the moment in which I write this text. What occurs in the atoms of your brain is not any more independent from what is happening in the atoms of mine: we communicate.

The world isn't, then, just a network of colliding atoms: it is also a network of correlations between sets of atoms, a network of real reciprocal information between physical systems.

In all of this, there is nothing idealistic or spiritual; it's nothing but an application of Shannon's idea that alternatives can be counted. All this is as much a part of the world as the stones of the Dolomites, the buzzing of bees and the waves of the sea.

Once we have understood that this network of reciprocal information exists in the universe, it is natural to seek to use this treasure to describe the world. Let's start with an aspect of nature well understood since the end of the nineteenth century: heat. What *is* heat? What does it mean to say that something is hot? Why does a cup of scalding-hot tea cool itself down, rather than heating itself up further?

It was the Austrian scientist Ludwig Boltzmann, the founder of statistical mechanics, who first understood why.* Heat is the random microscopic movement of molecules: when the tea is hotter, the movement of the molecules is more agitated. Why does it cool down? Boltzmann hazarded a splendid hypothesis: because the number of possible states of the molecules in hot tea and cold air is smaller than the number in cool tea and slightly warmer air. The combined state evolves from a situation where there are less possible states to a situation where there are more possible states. The tea can't warm itself up, because information cannot increase by itself.

I'll elaborate. The molecules of tea are extremely numerous and extremely small, and we don't know their precise movements. Therefore, we lack information. This lack of information – or missing information – can be computed. (Boltzmann did it: he computed the number of distinct states

---

* Boltzmann did not use the concept of information, but his work can be read in this way.

the molecules can be in. This number depends on the temperature.) If the tea cools, a little of its energy passes into the surrounding air; therefore, the molecules of tea move more slowly and the molecules of air move more quickly. If you compute your missing information, you discover that it has increased. If, instead, tea absorbed heat from the colder air, then the missing information would be decreased. That is, we would know more. But information cannot fall from the sky. It cannot increase by itself, because what we don't know, we just don't know. Therefore, the tea cannot warm up by itself in contact with cold air. It sounds a bit magical, but it works: we can predict how heat behaves just on the basis of the observation that our information cannot increase for free!

Boltzmann was not taken seriously. At the age of fifty-six, in Duino, near Trieste, he committed suicide. Today, he is considered one of the geniuses of physics. His tomb is incised with his formula

$$S = k \log W$$

which expresses (missing) information as the logarithm of the number of alternatives, Shannon's key idea. Boltzmann pointed out that this quantity coincides with the entropy used in thermodynamics. Entropy is 'missing information', that is, information with a minus sign. The total amount of entropy can only increase, because information can only diminish.*

---

* Entropy is proportional to the logarithm of the volume of the phase space. The constant of proportionality, k, is Boltzmann's constant, which transforms the units of measurement for information (bits) into the units of measurement for entropy (Kelvin's joules).

Today, physicists commonly accept the idea that information can be used as a conceptual tool to throw light on the nature of heat. More audacious, but defended today by an increasing number of theorists, is the idea that the concept of information can be useful also to the mysterious aspects of quantum mechanics illustrated in Chapter 5.

Remember that a key result of quantum mechanics is precisely the fact that information is finite. The number of alternative results that we can obtain measuring a physical system* is infinite in classical mechanics; but, thanks to quantum theory, we have understood that, in reality, it is finite. Quantum mechanics can be understood as the discovery that information in nature is always *finite*.

In fact, the entire structure of quantum mechanics can be read and understood in terms of information, as follows. A physical system manifests itself only by interacting with another. The description of a physical system, then, is always given in relation to another physical system, the one with which it interacts. Any description of a system is therefore always a description of the *information* which a system has about another system, that is to say, the *correlation* between the two systems. The mysteries of quantum mechanics become less dense if interpreted in this way, as the description of the information that physical systems have about one another.

The description of a system, in the end, is nothing other than a way of summarizing all the past interactions with it, and using them to predict the effect of future interactions.

The entire formal structure of quantum mechanics can in large measure be expressed in two simple postulates:[1]

* In a finite region of its phase space.

1. The relevant information in any physical system is finite.
2. You can always obtain new information on a physical system.

Here, the 'relevant information' is the information that we have about a given *system* as a consequence of our past interactions with it: information allowing us to predict what will be the result for us of future interactions with this system. The first postulate characterizes the granularity of quantum mechanics: the fact that a finite number of possibilities exists. The second characterizes its indeterminacy: the fact that there is always something unpredictable which allows us to obtain *new* information. When we acquire new information about a system, the total relevant information cannot grow indefinitely (because of the first postulate), and part of the previous information becomes *irrelevant*, that is to say, it no longer has any effect upon predictions of the future. In quantum mechanics when we interact with a system, we don't only learn something, we also 'cancel' a part of the relevant information about the system.*

The entire formal structure of quantum mechanics follows in large measure from these two simple postulates. Therefore, the theory lends itself in a surprising way to being expressed in terms of information.

The first to realize that the notion of information was fundamental to the understanding of quantum reality was John Wheeler, the father of quantum gravity. Wheeler coined the phrase 'It from bit' to express this idea, meaning that 'everything is information'.

* This is what came to be called, inappropriately, the 'collapse' of the wave function.

Information reappears, then, in the context of quantum gravity. Remember: the area of any surface is determined by the spins of the loop which intersect this surface. These spins are discrete quantities, and each one contributes to the area.

A surface with a fixed area may be formed from these elementary quanta of area in many different ways, say, in a number of ways N. If you know the area of a surface but don't know exactly how its quanta of area are distributed, you have missing information about the surface. This is one of the ways of computing the heat of black holes: the quanta of area of a black hole enclosed in a surface of a certain area can be in N different possible distributions. It is like for the cup of tea, in which the molecules can move in N different possible ways. Thus we can associate a quantity of missing information, that is to say, entropy, with a black hole.

The amount of information associated thus with a black hole depends directly upon the area A of the hole: the larger the hole, the greater the amount of missing information.

When information enters into a black hole, it is no longer recoverable from outside. But the information which enters the black hole carries with it the energy by which the black hole becomes larger and increases its area. Viewed from outside, the information lost in the black hole now appears as entropy associated with the area of the hole. The first to suspect something similar was the Israeli physicist Jacob Bekenstein.

But the situation is anything but clear because, as we have seen in the last chapter, black holes emit thermal radiation and very slowly evaporate, becoming smaller and smaller until they probably disappear, subsumed in that ocean of microscopic black holes which constitutes space at the

Planck scale. Where does the information that has fallen into the black hole as the black hole shrinks end up? Theoretical physicists are debating the question, and no one has a completely clear answer.

All of this, I believe, indicates that in order to grasp the basic grammar of the world, we need to merge three basic ingredients, not just two: not just general relativity and quantum mechanics, but also the theory of heat, that is, statistical mechanics and thermodynamics, which we can also describe as information theory. But the thermodynamics of general relativity, that is to say, the statistical mechanics of quanta of space, is as yet only in its first infancy. Everything is still confused, and there is a very great deal which remains to be understood.

All of this brings us to the last idea I describe in this book: thermal time.

Thermal time

The problem at the root of the idea of thermal time is simple. In Chapter 7, I showed that it is not necessary to use the notion of time to describe physics. It is better to forget time altogether. Time plays no role at the fundamental level of physics. Once we have understood this, it is easier to write the equations of quantum gravity.

There are many everyday notions which no longer have any role in the fundamental equations of the universe; for example, the notions of 'up' and 'down', or 'hot' and 'cold', so it is not particularly strange that shared quotidian notions disappear from fundamental physics. However, once we have accepted this idea, we obviously open up a second problem.

How can we recover the notions of our everyday experience? How do they emerge, in our specific context?

For example, the notions of 'up' and 'down' don't enter into Newton's equations, but we know what they mean in a schema without absolute up and down. 'Up' and 'down' are meaningful near a large mass, like a planet. 'Down' indicates the direction towards which the large, near mass exerts gravitational pull; 'up' indicates the opposite direction. The same goes for 'hot' and 'cold': there are no 'hot' or 'cold' things at a microscopic level but, when we put together a large number of microscopic constituents and describe them in terms of averages, then the notion of 'heat' appears: a hot body is a body where the average speed of single constituents is raised. We are able to understand the meaning of 'up' or 'hot' in certain situations: the presence of a nearby mass, or the fact that we are dealing only with average values of many molecules, and so on.

Something similar must apply to 'time'. If the notion of time has no role to play at an elementary level, it certainly plays a significant role in our lives, just as 'up' and 'hot' do. What does 'the passage of time' mean, if time plays no part in the fundamental description of the world?

The answer is simple. The origin of time may be similar to that of heat: it comes from averages of many microscopic variables. Let's look at this in detail.

That there is a link between time and temperature is an ancient and recurrent idea. If you think about it, all phenomena where we detect the passage of time are co-involved with temperature. The salient characteristic of time is that it moves forwards and not backwards, that is to say, there are *irreversible phenomena*. Mechanical phenomena – ones that don't involve heat – are reversible. If we film them and then run the film backwards, we see something realistic. If we film a swinging pendulum, or a stone thrown upwards then falling, and then watch the film in reverse, we still see a plausible pendulum swinging, or a stone rising and dropping to the ground.

When the stone reaches the ground, it stops, you might object: if you watch the film reversed, you see a stone leaping up from the ground by itself, and this is implausible. But when the stone reaches the ground and stops, where does its energy go? It *heats* the ground! At the precise moment when heat is produced, the process is irreversible: the past differs from the future. It is always heat and only heat that distinguishes the past from the future.

This is universal. A burning candle is transformed into smoke – the smoke cannot transform into a candle – and a candle produces heat. A boiling-hot cup of tea cools down and does not heat up: it diffuses heat. We live and get old: producing heat. Our old bicycle wears out with time: producing heat through friction. Think of the solar system. At first approximation, it continues to turn like an immense mechanism always equal to itself. It doesn't produce heat and, in fact, if you watched it in reverse you wouldn't notice anything strange about it. But looked at more closely, there are also irreversible phenomena: the Sun is using up its combustible hydrogen and will eventually exhaust it and extinguish: the Sun, too, is getting older and, in fact, produces heat. The Moon also appears to orbit the Earth unchangingly and be always equal to itself, whereas in reality it is slowly moving away. This is because it raises tides, and the tides heat the sea a little, thus exchanging energy with the Moon. Whenever you consider a phenomenon certifying the passage of time, it is through the production of heat that it does so. There is no preferred direction of time without heat.

But heat is our way to name averages over many variables.

The idea of thermal time reverses this observation. That is to say, instead of enquiring how time produces dissipation in heat, it asks how heat produces time.

Thanks to Boltzmann, we know that the notion of heat comes from the fact that we interact with averages. The idea of thermal time is that the notion of time, too, comes from the fact that we interact only with averages of many variables.*

As long as we have a *complete* description of a system, all the variables of the system are on the same footing; none of them acts as a time variable. That is to say: none is correlated to irreversible phenomena. But as soon as we describe the system by means of averages of many variables, we have a preferred variable that functions like common time. A time along which heat is dissipated. The time of our everyday experience.

Hence time is not a fundamental constituent of the world, but it appears because the world is immense, and we are small systems within the world, interacting only with macroscopic variables that average among innumerable small, microscopic variables. We, in our everyday lives, never see a single elementary particle, or a single quantum of space. We see stones,

* Here is how it works technically: a Boltzmann statistical state is described by a function on phase space given by the exponential of the Hamiltonian. The Hamiltonian is the generator of time evolution. In a system in which time is not defined, there is no Hamiltonian. But if we have a statistical state, we just take its logarithm and this defines a Hamiltonian, and hence a notion of time.

mountains, the faces of our friends – and each of these things we see is formed by myriads of elementary components. We are always correlated with averages. Averages behave like averages: they disperse heat and, intrinsically, generate time.

The difficulty of grasping this idea comes from the fact that it is hard for us to think of a world without time, and of time emerging in an approximate manner. We are too used to thinking of reality as existing in time. We are beings who live in time: we dwell in time, and are nourished by it. We are an effect of this temporality, produced by average values of microscopic variables. But the limitations of our intuitions should not mislead us. Understanding the world better often entails going against intuition. If this were not the case, understanding would be easy.

Time is an effect of our overlooking of the physical microstates of things. Time is information we don't have.

Time is our ignorance.

Reality and information

Why does information play such a central role as this? Perhaps because we must not confuse what we know about a system with the absolute state of the same system. What we know is something concerning the relation between the system and ourselves. Knowledge is intrinsically relational; it depends just as much on its object as upon its subject. The notion of the 'state' of a system refers, explicitly or implicitly, to another system. Classical mechanics misled us into thinking that we could do without taking account of this simple truth, and that we could access, at least in theory, a vision of reality entirely independent of the observer. But the development of physics has shown that, at the end of the day, this is impossible.

Careful: when I say that we 'have information' about the temperature of cup of tea, or we 'don't have information' about the velocity of every single molecule, I am not saying something about mental states, or abstract ideas. I am only saying that the laws of physics determine a correlation between ourselves and the temperature (for instance, I've looked at a thermometer), but not between ourselves and the velocity of the individual molecules. It is the same notion of information as the one I started from in this chapter: the white ball in your hand 'has information' about the fact that the ball in my hand is black. We're dealing with physical facts, not mental notions. A ball has information, in this sense, even if the ball does not have mental states, just as a USB storage device contains information (the number of giga-bytes printed on the device tells us how much information it can contain), even if a USB storage device does not think. Information in this sense – correlation between states of systems – is ubiquitous throughout the universe.

I believe that in order to understand reality we have to keep in mind that reality is this network of relations, of reciprocal information, which weaves the world. We slice up the reality surrounding us into objects. But reality is not made up of discrete objects. It is a variable flux. Think of an ocean wave. Where does a wave finish? Where does it begin? Think of mountains. Where does a mountain start? Where does it end? How far does it continue beneath the Earth's surface? These are questions without much sense, because a wave and a mountain are not objects in themselves; they are ways which *we* have of slicing up the world to apprehend it, to speak about it more easily. These limits are arbitrary, conventional, comfortable: they depend on us (as physical systems) more than on the waves or the mountains. They are ways of organizing the information which we have or, better, forms of information which we have.

It's the same for every object, properly considered, including living organisms. This is why it makes little sense to ask whether a half-cut fingernail is still 'me' or has become 'not-me'; or if the hairs left on my sofa by the cat are still part of the cat, or not; or precisely when a child's life begins. A child begins to live on the day when a person dreams of her for the first time, long before her conception, or when she forms her first self-image, or when she breathes for the first time, or when she recognizes her name, or when we apply any number of other conventions: they are all useful, but arbitrary. They are ways to think, and to orientate ourselves within the complexity of reality.

A living organism is a system which continually re-forms itself in order to remain itself, interacting ceaselessly with the external world. Of such organisms, only those continue to exist which are more efficient at doing so and, therefore, living organisms manifest properties which have suited them for survival. For this reason, they are interpretable, and we interpret them, in terms of intentionality, of purpose. The finalistic aspects of the biological world (this is Darwin's momentous discovery) are therefore the result of the selection of complex forms effective in persisting. But the effective way of continuing to exist in a changing environment is to manage correlations with the external world better, that is to say, information; to collect, store, transmit and elaborate information. For this reason, DNA exists, together with immune systems, sense organs, nervous systems, complex brains, languages, books, the library of Alexandria, computers and Wikipedia: they maximize the efficiency of information management – the management of correlations favouring survival.

The statue that Aristotle sees in a block of marble is more than the block of marble: but it is not an abstract form that resides just in the statue. It is something residing in the correlations between the mind of Aristotle, or ours, and the marble; something that pertains to the information which the marble provides regarding something that is significant for Aristotle, or for us. It is something regarding a discus thrower, Phidias, Aristotle and the marble, and resides in the correlated dispositions of the atoms of the statue, and the correlations between these and a thousand others, in our minds or in Aristotle's. These speak of a discus thrower, just as the white ball in your hand tells you that the ball in mine is black. We are structured to manage precisely this – information – and remain in existence thanks to this.

Even from this brief overview it should be clear that the notion of information plays a central role in our attempts to understand the world. From communication to the basis of genetics, from thermodynamics to quantum mechanics and up to quantum gravity, the notion of information is gaining ground as a tool for understanding. The world should not be understood as an amorphous ensemble of atoms – but rather as a game of mirrors, founded on the correlations between the structures formed by *combinations* of these atoms.

As Democritus said, it is not just a question of these atoms but also of the *order* in which they are arranged. Atoms are like the letters in an alphabet: an extraordinary alphabet, so rich as to be able to read, reflect and even think about itself. We are not atoms; we are *orders* in which atoms are arranged, capable of mirroring other atoms and mirroring ourselves.

Democritus gave a strange definition of 'man': 'Man is

what we all know.'[2] At first sight, this seems rather silly and empty, but it is not so.

Salomon Luria, the major scholar of Democritus, observes that it is not a banality that Democritus is giving us. The nature of a man is not his internal structure but the network of personal, familial and social interactions within which he exists. It is these which 'make' us, these which guard us. As humans, we are that which others know of us, that which we know of ourselves, and that which others know about our knowledge. We are complex nodes in a rich web of reciprocal information.

All of this is not yet a theory. These are tracks we are following, I believe, in seeking to understand the world around us better. There still remains a great deal to understand. I'll speak of this in the final chapter.

13. Mystery

The truth is in the depths.
— Democritus[1]

I've described what I think is the nature of things in the light of what we have learned to date. I've summarized the development of some key ideas of fundamental physics, and I have illustrated the great discoveries made by physics in the twentieth century and the image of the world emerging from the research into a quantum theory of gravity.

Am I sure about all this? I am not.

One of the very first and most beautiful pages in the history of science is the passage in Plato's *Phaedo* in which Socrates explains the shape of the Earth.

Socrates says he 'believes' the Earth is a sphere, with great valleys where men live. He's basically right, if a bit confused. He adds, 'I'm not sure.' This page is worth much more than all of the nonsense on the immortality of the soul which fills the rest of the dialogue. It is not just the oldest text to come down to us which speaks explicitly of the fact that the Earth must be spherical. More importantly, it shines with the crystalline clarity with which Plato acknowledges the *limits* of the knowledge of his time. 'I'm not sure,' says Socrates.

This acute awareness of our ignorance is the heart of scientific thinking. It is thanks to this awareness of the limits of our knowledge that we have learned so much. We are not certain of all which we suspect, just as Socrates was not sure of the spherical nature of the Earth. We are exploring at the borders of our knowledge.

Awareness of the limits of our knowledge is also awareness of the fact that what we know may turn out to be wrong, or inexact. Only by keeping in mind that our beliefs may turn out to be wrong is it possible to free ourselves from wrong ideas, and to learn. To learn something, it is necessary to have the courage to accept that what we think we know, including our most rooted convictions, may be wrong, or at least naïve: shadows on the walls of Plato's cave.

Science is born from this act of humility: not trusting blindly in our past knowledge and our intuition. Not believing what everyone says. Not having faith in the accumulated knowledge of our fathers and grandfathers. We learn nothing if we think that we already know the essentials, if we assume that they were written in a book or known by the elders of the tribe. The centuries in which people had faith in what they believed were the centuries in which little new was learned. Had they trusted the knowledge of their fathers, Einstein, Newton and Copernicus would never have called things into question and would have never been able to move our knowledge forwards. If no one had raised doubts, we would be still worshipping pharaohs and thinking that the Earth is supported on the back of a giant turtle. Even our most efficacious knowledge, such as that found by Newton, may eventually turn out, as Einstein showed, to be simplistic.

Science is sometimes criticized for pretending to explain everything, for thinking that it has an answer to every question. It's a curious accusation. As every researcher working in every laboratory throughout the world knows, doing science means coming up hard against the limits of your ignorance on a daily basis – the innumerable things which you don't know, and can't do. This is quite different from claiming to know everything. We don't know which particles we might see next year at CERN, or what our next telescopes will reveal, or which equations truly describe the world; we don't know how to solve the equations we have, and sometimes we don't understand what they signify; we don't know if the beautiful theory on which we are working is right. We don't know what there is beyond the Big Bang; we don't know how a storm works, or a bacterium, or an eye – or the cells in our own bodies, or our thought processes. A scientist is someone who lives immersed in the awareness of our deep ignorance, in direct contact with our own innumerable limits, with the limits of our understanding.

But if we are certain of nothing, how can we possibly rely on what science tells us? The answer is simple. Science is not reliable because it provides certainty. It is reliable because it provides us with the best answers we have at present. Science is the most we know so far about the problems confronting us. It is precisely its openness, the fact that it constantly calls current knowledge into question, which guarantees that the answers it offers are the best so far available: if you find better answers, these new answers become science. When Einstein found better answers than Newton, he didn't question the capacity of science to give the best possible answers – on the contrary, he confirmed it.

The answers given by science, then, are not reliable because they are definitive. They are reliable because they are not definitive. They are reliable because they are the best available today. And they are the best we have because we don't consider them to be definitive, but see them as open to improvement. It's the awareness of our ignorance that gives science its reliability.

And it is reliability that we need, not certainty. We don't have absolute certainty, and never will have it – unless we accept blind belief. The most credible answers are the ones given by science, because science *is* the search for the most credible answers available, not for answers pretending to certainty.

Though rooted in previous knowledge, science is an adventure based on continuous change. The story I have told reaches back over millennia, tracing a narrative of science that has treasured good ideas but hasn't hesitated to throw ideas away when something which works better was found. The nature of scientific thinking is critical, rebellious and dissatisfied with a priori conceptions, with reverence and sacred or untouchable truth. The search for knowledge is not nourished by certainty: it is nourished by a radical distrust in certainty.

This means not giving credence to those who say they are in possession of the truth. For this reason, science and religion frequently find themselves on a collision course. Not because science pretends to know ultimate answers but precisely for the opposite reason: because the scientific spirit distrusts whoever claims to be the one *having* ultimate answers, or privileged access to Truth. This distrust is found to be disturbing in some religious quarters. It is not science which is disturbed by religion: there are certain religions that are disturbed by scientific thinking.

To accept the substantial uncertainty of our knowledge is to accept living immersed in ignorance and, therefore, in mystery, to accept living with questions to which we do not know the answers. Perhaps we don't know them yet or – who knows? – we never will.

To live with uncertainty may be difficult. There are those

who prefer any certainty, even if unfounded, to the uncertainty which comes from recognizing our own limits. There are some who prefer to believe in a story just because it was believed by the tribe's ancestors rather than bravely to accept uncertainty.

Ignorance can be scary. Out of fear, we can tell ourselves calming stories: up there beyond the stars, there is an enchanted garden, with a gentle father who will welcome us into his arms. It doesn't matter if this is true, it is reassuring.

There is always, in this world, someone who pretends to tell us the ultimate answers. The world is full of people who say that they have The Truth. Because they have got it from the fathers; they have read it in a Great Book; they have received it directly from a god; they have found it in the depths of themselves. There is always someone who has the presumption to be the depository of Truth, neglecting to notice that the world is full of *other* depositories of Truth, each one with his own real Truth, different from that of the others. There is always some prophet dressed in white, uttering the words, 'Follow me, I am the true way.'

I don't criticize those who prefer to believe in this: we are all free to believe in whatever we want. Maybe, after all, there is a grain of truth in the joke reported by St Augustine: What was God doing before creating the world? He was preparing Hell for those who seek to scrutinize deep mysteries.[2] But these deep mysteries are precisely the 'depths' in which Democritus, in the quote that opens this chapter, invites us to seek the truth.

For my part, I prefer to look our ignorance in the face, accept it and seek to look just a bit further: to try to understand that which we are able to understand. Not just because accepting this ignorance is the way to avoid being entangled in superstitions and prejudices – but because to accept our ignorance in the first place seems to me to be the truest, the most beautiful and, above all, the most honest way.

To seek to look further, to go further, seems to me to be one of the splendid things which gives sense to life. Like loving, or looking at the sky. The curiosity to learn, to discover, to look over the next hill, the desire to taste the apple: these are the things which make us human. As Dante's Ulysses reminds his companions, we are not made 'to live like brutes, but to seek virtue and knowledge'.

The world is more extraordinary and profound than any of the fables told by our forefathers. I want to go and see it. To accept uncertainty doesn't detract from our sense of mystery. On the contrary: we are immersed in the mystery and the beauty of the world. The world revealed by quantum gravity is a new and strange one – still full of mystery, but coherent with its simple and clear beauty.

It is a world which does not exist in space and does not develop in time. A world made up solely of interacting quantum fields the swarming of which generates – through a dense network of reciprocal interactions – space, time, particles, waves and light (figure 13.1)

> It continues, it continues, teeming life, and death
> Tender and hostile, clear and unknowable.

And the poet goes on:

> So much the eye can see, from this watching tower.[3]

A world without infinity, where the infinitely small does not exist, because there is a minimum scale to this teeming,

Figure 13.1 An intuitive representation of quantum gravity.

beneath which there is nothing. Quanta of space mingle with the foam of spacetime, and the structure of things is born from reciprocal information which weaves the correlations between the regions of the world. A world which we know how to describe with a set of equations. Perhaps, to be corrected.

It's a vast world, with much still to clarify and explore. It's my fondest dream that someone – one of the younger readers of this book, I hope – will be able to voyage across it and illuminate it better. Beyond the next hill there are worlds still more vast, still to be discovered.

Annotated Bibliography

Andolfo, Matteo, *Atomisti antichi. Frammenti e testimonianze* (*Ancient Atomism. Fragments and Testominies*), Milan, Rusconi, 1999.

Aristotle, *On Generation and Corruption,* in *The Complete Works of Aristotle,* Vol. I. Ed. Jonathan Barnes, Princeton, Princeton University Press, 1984. The principal work by Aristotle which provides information on the thought of Democritus.

Bitbol, Michel, *Physical Relations or Functional Relations? A Non-metaphysical Construal of Rovelli's Quantum Mechanics.* Philosophy of Science Archives, 2007, http://philsci-archive.pitt.edu/3506/. Commentary on and Kantian interpretation of relational quantum mechanics.

Baggott, Jim, *The Quantum Story: A History in Forty Moments.* New York, Oxford University Press, 2011. A good and complete reconstruction of the main stages in the development of quantum mechanics, from its origin to the present day.

Bojowald, Martin, *Once Before Time: A Whole Story of the Universe.* New York, Alfred A. Knopf, 2010. An explanatory description of the application of loop quantum gravity to the origin of the universe – by one of the first scientists to use it in this way. Includes an account of the so-called 'Big Bounce' of the universe, which may have happened before the Big Bang.

Calaprice, Alice (ed.), *Dear Professor Einstein. Albert Einstein's Letters to and from Children.* New York, Prometheus Books, 2002. A delightful collection of letters exchanged between Einstein and several children.

Democritus, *Raccolta dei frammenti* (*Collected Fragments*). Interpretation and Commentary by S. Luria. Trans. It. Milan, Bompiani, 2007. There is no English translation of this classic edition.

Democritus, *The Atomists: Leucippus and Democritus. Fragments*. A text and translation with commentary by C. C. W. Taylor. Toronto, Toronto University Press, 1999. Diehls, H. and Kranz, W. (eds), *Die Fragmente der Vorsokratiker*. Berlin, Weidmann, 1903. The classic collection, in three volumes, of the 'fragments' of the Presocratics.

Dorato, Mauro, *Che cos'è il tempo? Einstein, Gödel e l'esperienza commune*. Rome, Carocci, 2013. Precise and complete discussion, centred around special relativity, of the Einsteinian modification of the concept of time.

—, *Rovelli's Relational Quantum Mechanics, Monism and Quantum Becoming*. Philosophy of Science Archives, 2013, http://philsci-archive.pitt.edu/9964/. Discussion by an Italian philosopher of interpretations of quantum mechanics.

Fano, Vincenzo, *I paradossi di Zenone*. Trans. It. Rome, Carocci, 2012. An excellent study which highlights the topicality of the problems posed by Zeno's paradoxes.

Farmelo, Graham, *The Strangest Man: The Hidden Life of Paul Dirac, Quantum Genius*. London, Faber, 2009. An extensive but highly readable account of the life and baffling character of the greatest physicist after Einstein.

Feynman, Richard, *The Feynman Lectures on Physics*, eds. Richard B. Leighton and Matthew Sands (3 vols.). London, Basic Books, 2011. Essential physics textbook drawn from the lectures of the greatest American physicist. They are brilliant, original, lively, the product of an outstanding intelligence. No student of physics who is truly interested in science should fail to read them.

Fölsing, Albrecht, *Albert Einstein: A Biography*. New York, Penguin, 1998. Extensive and complete biography of Einstein.

Gorelik, Gennaly E. and Frenkel, Victor, *Matvei Petrovich Bronstein and Soviet Theoretical Physics in the Thirties*. Boston, Birkhauser Verlag, 1994. Historical study of Bronštejn, the young Russian who initiated research in quantum gravity and was executed by Stalin.

Greenblatt, Stephen, *The Swerve: How the World became Modern*. New York, W. W. Norton, 2011. A book which reconstructs the influence of the rediscovery of Lucretius upon the birth of the modern world.

Heisenberg, Werner, *Physics and Philosophy*: *The Revolution in Modern Science*. New York, Harper & Row, 1962. The true originator of quantum mechanics reflects on general problems of philosophy and science.

Kumar, Manjit, *Quantum: Einstein, Bohr and the Great Debate about the Nature of Reality*. London, Icon Books, 2009. A fine explanatory, detailed reconstruction of the birth of quantum mechanics and, above all, of the long dialogue between Bohr and Einstein on the meaning of the new theory.

Lucretius, *On The Nature of the Universe* (*De rerum natura*), trans. R. E. Latham. Harmondsworth, Penguin, 1951. The wonderful poem from which we know the ideas and spirit of ancient atomism.

Newton, Isaac, *A Treatise of the System of the World*. London, F. Fayram, 1731. A little-known work by Newton in which he gives his theory of universal gravity in a much less technical form than in his great treatise (*Principia*).

Odifreddi, P., *Come stanno le cose. Il mio Lucrezio, la mia Venere*. Milan, Rizzoli, 2013. Translation with extensive commentary of Lucretius's poem. An ideal school textbook.

Plato, *Phaedo*, ed. David Gallop. Oxford, Oxford University Press, 2009. The oldest extant text to speak explicitly about the spherical nature of the earth.

Rovelli, Carlo, 'Aristotle's Physics: A Physicist's Look' in *Journal of the American Philosophical Association*, 1 (2015) 23–40.

—, *Seven Brief Lessons on Physics*, trans. Simon Carnell and Erica Segre. London, Allen Lane, 2015. A concise guide through the scientific revolution that shook physics in the twentieth century and still continues to shake us today.

—, *The First Scientist: Anaximander and His Legacy*, trans. Marion Lignana Rosenberg. Yardley, Westholme, 2007. A reconstruction of the ideas of Anaximander – one of the greatest scientists of all time – and their influence on the development of science. The book is also a reflection on the birth and nature of scientific thinking: its distinctive character, its difference from mythical thinking, its limitations and strength.

—, 'Quantum Gravity', in Butterfield, J. and Earman, J. (eds.), *Handbook of The Philosophy of Science, Philosophy of Physics*. Amsterdam, Elsevier/North-Holland, 2007, pp. 1287–330. A long article aimed at philosophers, with a detailed discussion of the current state of quantum gravity, of its unresolved problems and the various approaches to them.

—, *Quantum Gravity*. Cambridge, Cambridge University Press, 2004. Technical manual on quantum gravity. Definitely not recommended for anyone without a background in physics.

—, 'Relational Quantum Mechanics' in *The Stanford Encyclopaedia of Philosophy*, http://plato.stanford.edu/archives/win2003/entries/rovelli/. Synthesis of the relational interpretation of quantum mechanics.

—, 'Relational Quantum Mechanics' in *International Journal of Theoretical Physics*, 35 (1996), 1637, http://arxiv.org/abs/quant-ph/9609002. The article which first introduces the theory of relational quantum mechanics.

Smolin, Lee, *Three Roads to Quantum Gravity*. New York, Basic Books, 2002. An introduction to quantum gravity and its open questions.

Van Fraassen, Bas. 'Rovelli's World' in *Foundations of Physics*, 40 (2010), 390–417. A discussion of relational quantum mechanics, by an important analytic philosopher.

Notes

1. Grains

1 On Anaximander and the Milesians, see Carlo Rovelli, *The First Scientist: Anaximander and His Legacy*, trans. Marion Lignana Rosenberg (Yardley, Westholme, 2007).

2 The Milesian origin of Leucippus is given, for instance, by Simplicius (see M. Andolfo, *Atomisti antichi. Frammenti e testimonianze (Ancient Atomism. Fragments and Testimonies)*, (Milan, Rusconi, 1999), p. 103. But it is not certain. The reference to Miletus and to Elea is significant in relation to his cultural roots; the debt Leucippus owed to Zeno of Elea is discussed in the following pages.

3 Seneca, *Naturales questiones*, VII, 3, 2d.

4 Cicero, *Academica priora*, II, 23, 73.

5 Sextus Empiricus, *Adversus mathematicos*, VIII, 135 (trans. R. G. Bury *Against the Professors*), (Loeb Classical Library, 1989).

6 See Aristotle, *On Generation and Corruption*, A1, 315b 6, in *The Complete Works of Aristotle*, Vol. I, ed. Jonathan Barnes (Princeton, Princeton University Press, 1984).

7 A collection of ancient fragments and testimonies which speak of atomism is given in M. Andolfo's *Ancient Atomists*. A complete anthology of fragments and testimonies concerning Democritus has been published by Solomon Luria. (See entry 'Democritus' in the bibliography for an English alternative.)

8 For a brief and interesting recent work on the thought of Democritus, placing it in the context of humanism, see S. Martini,

Democrito: filosofo della natura o filosofo dell'uomo? (*Democritus: Philosopher of Nature or Philosopher of Man?*), (Rome, Armando, 2002.)

9 Plato, *Phaedo*, ed. David Gallup (Oxford, Oxford University Press, 2009, XLVI.

10 Richard Feynman, *The Feynman Lectures on Physics*, Vol. 1, eds. Robert Leighton and Matthew Sands (London, Basic Books, 2011).

11 See Aristotle, *On Generation and Decay*, A2, 316a, in *The Complete Works of Aristotle*, Vol. I, ed. Jonathan Barnes (Princeton, Princeton University Press, 1984).

12 A good recent account of Zeno's paradoxes, and of their philosophical and mathematical relevance, is given by Vincenzo Fano in *I paradossi di Zenone* (*Zeno's Paradoxes*), (Rome, Carocci, 2012).

13 *Amores* (Love Poems), I, 15, 23–4.

14 Lucretius, *On the Nature of the Universe*, trans. E. A. Latham, (Harmondsworth, Penguin, 1951), p. 173

15 Ibid., p. 89.

16 Ibid., p. 27

17 Ibid., p. 60.

18 Guido Cavalcanti, *Rime.*, trans. Leonard Cottrell.

19 For an account of the rediscovery of Lucretius's text and its impact upon European culture, see Stephen Greenblatt, *The Swerve: How the World Became Modern* (New York, Norton, 2011).

20 See M. Camarota, 'Galileo, Lucretius and Atomism', in F. Citti and M. Beretta (eds.), *Lucrezio, la natura e la scienza* (*Lucretius, Nature and Science*), (Florence, Leo S. Olshki, 2008), pp. 141–75.

21 See R. Kargon, *Atomism in England from Hariot to Newton* (Oxford, Oxford University Press, 1966).

22 William Shakespeare, *Romeo and Juliet*, 1.4, 55–60, *Complete Works*, eds. Jonathan Bate and Eric Rassmussen (London, Macmillan/The Royal Shakespeare Company, 2007), p. 1690.

23 *On The Nature of the Universe*, pp. 63–4.
24 Piergiorgio Odifreddi has published a fine translation of and commentary on Lucretius's text, designed for use in schools. (*Come stanno le cose. Il mio Lucrezio, la mia Venere (The Nature of Things. My Lucretius, My Venus)*, (Milan, Rizzoli, 2013). It would be wonderful if schools adopted this book and if this extraordinary text was more widely known. A reading of the text, and of its author, diametrically opposed to Odifreddi's is given by V. E. Alfieri in *Lucrezio (Lucretius)*, (Florence, Le Monnier, 1929) and emphasizes the poignant, poetic qualities of the work, deriving from them a noble but bitter interpretation of the character of Lucretius.
25 H. Diels and W. Kranz„ eds., *Die Fragmente der Vorsokratiker (The Presocratics: Fragments)*, (Berlin, Weidmann, 1903), 68 b 247.

2. The Classics

1 On the value of Aristotle's physics, see C. Rovelli, 'Aristotle's Physics: A Physicist's Look', in *Journal of the American Philosophical Association*, 1 (2015), pp. 23–40.
2 Iamblichus of Chalchis, *Summa pitagorica* (Milan, Bompiani, 2006). For an English translation, see Iamblichus, *Life of Pythagoras*, trans. Thomas Taylor (Vermont, Inner Traditions International, 1986).
3 Isaac Newton, *Opticks* (1704), (New York, Dover, 1979), p. 400.
4 Giacomo Leopardi, *The Canti, with a Selection of His Prose*, trans. J. G. Nichols (New York, Routledge, 2003), p. 53.
5 Isaac Newton, *Letters to Bentley* (Montana, Kessinger, 2010). Cited in H. S. Thayer, *Newton's Philosophy of Nature* (New York, Hafener, 1953), p. 54.
6 Ibid.

7 Michael Faraday, *Experimental Researches in Electricity* (London, Bernard Quaritch, 1839–55), 3 vols., pp. 436–7. 3. Albert

1 Simplicius, *Aristotelis Physica* (*Aristotle's Physics*), 28, 15.

2 A. Calaprice, *Dear Professor Einstein. Albert Einstein's Letters to and from Children* (New York, Prometheus Books, 2002), p. 140.

3 The letter is quoted in A. Fölsing, *Einstein: A Biography* (London, Penguin, 1998), p. 337.

4 F. P. De Ceglia (ed.) *Scienziati di Puglia: secoli V aC.-XXI, dC, Parte 3* (*Scientists of Puglia: From the Fifth Century BC to the Twentieth Centuries, Part 3*) (Bari, Adda, 2007), p. 18.

5 A. Calaprice, *Dear Professor Einstein. Albert Einstein's Letters to and from Children* (New York, Prometheus Books, 2002), p. 208.

4. Quanta

1 Albert Einstein, '*Über einen die Erzeugung und Verwandlung des Lichtes betreffenden Gesichtspunkt*', *Annalen der Physik*, 17, pp. 132–48.

2 For a biography of Dirac which illustrates his disconcerting personality, see Graham Farmelo, *The Strangest Man: The Hidden Life of Paul Dirac, Quantum Genius* (London, Faber, 2009).

3 Lucretius, *On the Nature of the Universe*, trans. E. A. Latham, (Harmondsworth, Penguin, 1951), II, 218.

4 For a discussion of this relational interpretation of quantum mechanics, see 'Relational Quantum Mechanics' in the online *Stanford Encyclopedia of Philosophy*, http://plato.stanford.edu/archives/win2003/entries/rovelli, or Carlo Rovelli, 'Relational Quantum Mechanics', in *International Journal of Theoretical Physics*, 35, 1637, 1996, http://arxiv.org/abs/quant-ph/9609002.

5 Bas van Frassen, 'Rovelli's World', in *Foundations of Physics*, 40, 2010, 390–417; Michel Bitbol, *Physical Relations or Functional Relations? A Non-metaphysical Construal of Rovelli's Relational*

Quantum Mechanics, Philosophy of Science Archives, 2007, http://philsci-archive.pitt.edu/3506/; Mauro Dorato, *Rovelli's Relational Quantum Mechanics, Monism and Quantum Becoming,* Philosophy of Science Archives, 2013, http://philsci-archive. pitt.edu/9964/, and *Che cos'è il tempo? Einstein, Gödel e l'esperienza commune* (*What is Time? Einstein, Gödel and Shared Experience*), (Rome, Carocci, 2013).

5. Spacetime is Quantum

1 The work on the measurability of fields by Niels Bohr and Leon Rosenfeld is '*Det Kongelike Danske Videnskabernes Selskabs*', in *Mathematiks-fysike Meddelelser,* 12, 1933.
2 See Matvei Bronštejn, '*Quantentheorie schwacher Gravitationsfelder*', in *Physikalische Zeitschrift der Sowjetunion,* 9, 1936, 140–57; and '*Kvantovanie gravitatsionnykh voln*', in *Pi'sma v Zhurnal Eksperimental'noi I Teoreticheskoi Fiziki,* 6, 1936, 195–236.
3 See F. Gorelik and V. Frenkel, *Matvei Petrovich Bronstein and Soviet Theoretical Physics* (Boston, Birkhauser Verlag, 1994). 'Bronstein' was also the real surname of Trotsky.
4 The episode is recalled by Bryce DeWitt, http://www.aip.org/ history/ohilist/23199.html.

7. Time Does Not Exist

1 Lucretius, *On the Nature of the Universe,* trans. E. A. Latham, (Harmondsworth, Penguin, 1951), p. 41.
2 William Shakespeare, *A Midsummer Night's Dream,* 5.1, 23–7, RSC edition, p. 403.

8. Beyond the Big Bang

1 The speech can be found on the Vatican website: http://www.vatican.va/holy_father/pius_xii/speeches/1951/documents/hf_p-xii_spe_19511122_di-serena_it.html#top.

2 See S. Singh, *Big Bang* (London, Harper Collins, 2010),p. 362.

12. Information

1 A detailed discussion of these two postulates can be found in Carlo Rovelli, 'Relational Quantum Mechanics', in *International Journal of Theoretical Physics*, 35, 1637, 1996, http://arxiv.org/abs/quant-ph/9609002.

2 Cicero, *Academica priora*, II, 23, 73.

13. Mystery

1 Cited in Diogenes Laertius, *Lives of Eminent Philosophers*, 2 vols., (New York, Loeb, 1989).

2 St Augustine, *Confessions*, XI, 12 (Harmondsworth, Penguin, 2002).

3 Mario Luzi, *Dalla torre*, in *Dal fondo delle campagne* (Turin, Einaudi) p. 214.

Index

Page references in *italic* indicate Figures.

amplitude of transition
  calculation 104
and black holes 197–201, *198*
collapse of the wave
  function 217n
conflict with general
  relativity theory 125–7
and cosmology 179–82,
  183–94
Dirac's 102–5, 107–8
and Einstein 92–5, 118–19
and electromagnetism 92,
  95, 107–9
and the finite possibilities of
  information 111–12, 204,
  204, 216
and granularity 95, 98, 104,
  107–12, 116–17, 126
and identity of fields and
  particles 107–10
indeterminacy *see*
  indeterminacy/probability,
  quantum mechanics
information and structure
  of 216–18
limitation to divisibility of
  space in general relativity
  theory with 129–30,
  140–47, 202, 203
and loop quantum gravity
  144–5 *see also* loop theory/
  loop quantum gravity
quanta of space 125–37,
  138–50

quantum clouds 104–5, 107,
  114, 123, 149, 161, 181
quantum leaps 98, 99,
  100–101
quantum repulsion 180, 200
relationality 100–101, 102–3,
  115–17, 120
resolution with general
  relativity theory 168–70
  see also quantum gravity
and the Schrödinger wave
  105n
and special relativity theory
  107, 108
spectrum of variable
  calculation 104
and standard model of
  elementary particles
  108–10, 163, 165, 188–90
technologies dependent
  on 105

relationality
  of knowledge 223
  in quantum mechanics
    100–101, 102–3, 115–17,
    120
relativity
  general relativity theory *see*
    general relativity theory
  of simultaneity 53–7, *56*, 115
  special relativity theory
    52–60
  of velocity 53